Dimensions Math®
Textbook 5A

Authors and Reviewers
Bill Jackson
Jenny Kempe
Cassandra Turner
Allison Coates
Tricia Salerno

Consultant
Dr. Richard Askey

Singapore Math Inc.

Published by Singapore Math Inc.

19535 SW 129th Avenue
Tualatin, OR 97062
www.singaporemath.com

Dimensions Math® Textbook 5A
ISBN 978-1-947226-12-8

First published 2020
Reprinted 2020 (twice), 2021 (twice), 2022

Copyright © 2017 by Singapore Math Inc.
All rights reserved. This book or any portion thereof may not be reproduced or used in any manner whatsoever without the express written permission of the publisher.

Printed in China

Acknowledgments

Editing by the Singapore Math Inc. team.
Design and illustration by Cameron Wray with Carli Bartlett.

Preface

The Dimensions Math® Pre-Kindergarten to Grade 5 series is based on the pedagogy and methodology of math education in Singapore. The curriculum develops concepts in increasing levels of abstraction, emphasizing the three pedagogical stages: Concrete, Pictorial, and Abstract. Each topic is introduced, then thoughtfully developed through the use of problem solving, student discourse, and opportunities for mastery of skills.

Features and Lesson Components

Students work through the lessons with the help of five friends: Emma, Alex, Sofia, Dion, and Mei. The characters appear throughout the series and help students develop metacognitive reasoning through questions, hints, and ideas.

The colored boxes ▊ and blank lines in the textbook lessons are used to facilitate student discussion. Rather than writing in the textbooks, students can use whiteboards or notebooks to record their ideas, methods, and solutions.

Chapter Opener

Each chapter begins with an engaging scenario that stimulates student curiosity in new concepts. This scenario also provides teachers an opportunity to review skills.

Think

Students, with guidance from teachers, solve a problem using a variety of methods.

Learn

One or more solutions to the problem in **Think** are presented, along with definitions and other information to consolidate the concepts introduced in **Think**.

Do

A variety of practice problems allow teachers to lead discussion or encourage independent mastery. These activities solidify and deepen student understanding of the concepts.

Exercise

A pencil icon ━━━━━▶ at the end of the lesson links to additional practice problems in the workbook.

Practice

Periodic practice provides teachers with opportunities for consolidation, remediation, and assessment.

Review

Cumulative reviews provide ongoing practice of concepts and skills.

Emma　　Alex　　Sofia　　Dion　　Mei

Contents

Chapter		Lesson	Page
Chapter 1 **Whole Numbers**		Chapter Opener	1
	1	Numbers to One Billion	2
	2	Multiplying by 10, 100, and 1,000	6
	3	Dividing by 10, 100, and 1,000	10
	4	Multiplying by Tens, Hundreds, and Thousands	14
	5	Dividing by Tens, Hundreds, and Thousands	18
	6	Practice	22
Chapter 2 **Writing and Evaluating Expressions**		Chapter Opener	23
	1	Expressions with Parentheses	24
	2	Order of Operations — Part 1	28
	3	Order of Operations — Part 2	32
	4	Other Ways to Write and Evaluate Expressions	34
	5	Word Problems — Part 1	38
	6	Word Problems — Part 2	41
	7	Practice	44

Chapter	Lesson	Page
Chapter 3 **Multiplication and Division**	Chapter Opener	47
	1 Multiplying by a 2-Digit Number — Part 1	48
	2 Multiplying by a 2-Digit Number — Part 2	52
	3 Practice A	56
	4 Dividing by a Multiple of Ten	57
	5 Divide a 2-Digit Number by a 2-Digit Number	60
	6 Divide a 3-Digit Number by a 2-Digit Number — Part 1	63
	7 Divide a 3-Digit Number by a 2-Digit Number — Part 2	66
	8 Divide a 4-Digit Number by a 2-Digit Number	70
	9 Practice B	74
Chapter 4 **Addition and Subtraction of Fractions**	Chapter Opener	75
	1 Fractions and Division	76
	2 Adding Unlike Fractions	81
	3 Subtracting Unlike Fractions	85
	4 Practice A	88
	5 Adding Mixed Numbers — Part 1	90
	6 Adding Mixed Numbers — Part 2	94
	7 Subtracting Mixed Numbers — Part 1	98
	8 Subtracting Mixed Numbers — Part 2	102
	9 Practice B	106
	Review 1	108

Chapter	Lesson	Page

Chapter 5
Multiplication of Fractions

	Chapter Opener	111
1	Multiplying a Fraction by a Whole Number	112
2	Multiplying a Whole Number by a Fraction	115
3	Word Problems — Part 1	120
4	Practice A	124
5	Multiplying a Fraction by a Unit Fraction	126
6	Multiplying a Fraction by a Fraction — Part 1	129
7	Multiplying a Fraction by a Fraction — Part 2	132
8	Multiplying Mixed Numbers	135
9	Word Problems — Part 2	138
10	Fractions and Reciprocals	142
11	Practice B	145

Chapter 6
Division of Fractions

	Chapter Opener	147
1	Dividing a Unit Fraction by a Whole Number	148
2	Dividing a Fraction by a Whole Number	151
3	Practice A	155
4	Dividing a Whole Number by a Unit Fraction	156
5	Dividing a Whole Number by a Fraction	159
6	Word Problems	163
7	Practice B	166

Chapter		Lesson	Page

Chapter 7
Measurement

		Chapter Opener	167
	1	Fractions and Measurement Conversions	168
	2	Fractions and Area	172
	3	Practice A	176
	4	Area of a Triangle — Part 1	178
	5	Area of a Triangle — Part 2	184
	6	Area of Complex Figures	189
	7	Practice B	193

Chapter 8
Volume of Solid Figures

		Chapter Opener	195
	1	Cubic Units	196
	2	Volume of Cuboids	200
	3	Finding the Length of an Edge	205
	4	Practice A	209
	5	Volume of Complex Shapes	212
	6	Volume and Capacity — Part 1	216
	7	Volume and Capacity — Part 2	220
	8	Practice B	223
		Review 2	225

Chapter 1

Whole Numbers

Lesson 1
Numbers to One Billion

Think

One day last spring, the sun was 149,597,871 km from Earth.

(a) How do we read this number?

(b) What is the value of each of the first three digits in 149,597,871?

Learn

(a)

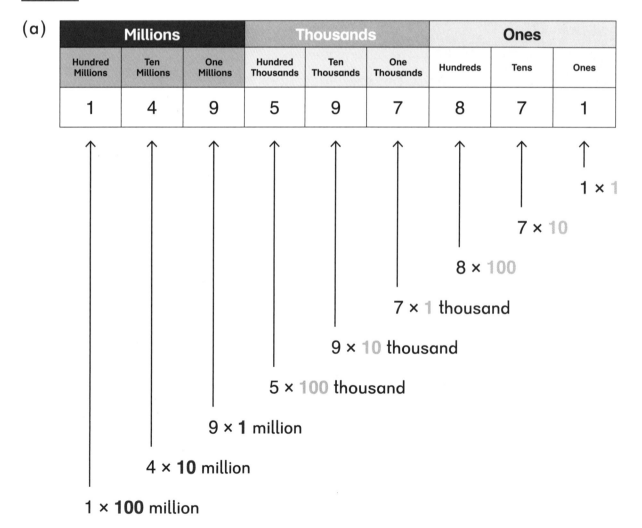

Every three places is called a **period**. We can place a comma in between each period to make large numbers easier to read.

1 4 9, 5 9 7 , 8 7 1

↑   ↑   ↑

149 million   597 thousand   871

One hundred forty-nine million, five hundred ninety-seven thousand, eight hundred seventy-one

(b) 149,597,871

The value of the digit 1 in the hundred millions place is:
1 × 100,000,000 = 100,000,000.

The value of the digit 4 in the ten millions place is:
4 × 10,000,000 = 40,000,000.

The value of the digit 9 in the one millions place is:
9 × 1,000,000 = 9,000,000.

Ten hundred millions is one **billion**. It is written as 1,000,000,000.

1-1 Numbers to One Billion

Do

1 (a) Read the number shown on the place-value chart.

Millions			Thousands			Ones		
Hundred Millions	Ten Millions	One Millions	Hundred Thousands	Ten Thousands	One Thousands	Hundreds	Tens	Ones
5	8	3	6	4	2	1	7	9

(b) What are the values of the digits 3, 8, and 5?

3 × 1,000,000 = ☐

8 × 10,000,000 = ☐

5 × 100,000,000 = ☐

2 Write each number in words.

(a) 25,604,067

(b) 372,805,005

(c) 650,700,802

(d) 909,021,014

3 (a) 2,000,000 + 600,000 + 50,000 + 30 = ☐

(b) 20,000,000 + 500,000 + 50,000 + 700 + 8 = ☐

(c) 7,000,000 + 465,000 + 652 = ☐

(d) 48,723,904 = ☐ + 723,000 + 904

1-1 Numbers to One Billion

4 (a) 1 million = ☐ hundred thousands

(b) 1 ten million = ☐ millions

(c) 1 hundred million = ☐ ten millions

(d) 200,000,000 = ☐ hundred millions

(e) 200,000,000 = ☐ millions

(f) 200,000,000 = ☐ thousands

5 Put the numbers in order from least to greatest.

| 765,308,205 | 665,308,205 | 99,999,999 | 745,308,205 |

6 Use the digits 0 to 8. Use each digit once.

(a) Write the greatest 9-digit odd number that has 7 in the ten millions place.

(b) Write the least 9-digit even number that has 5 in the ten millions place.

7 The number that is 1 more than 999,999,999 is one billion, which is written as 1,000,000,000. Write the following numbers in numerals.

(a) 5 billion (b) 25 billion (c) 370 billion

Exercise 1 • page 1

Lesson 2
Multiplying by 10, 100, and 1,000

Think

There are 231 posters in Box A.

(a) There are 10 times as many posters in Box B as in Box A. How many posters are in Box B?

(b) There are 100 times as many posters in Box C as in Box A. How many posters are in Box C?

(c) There are 1,000 times as many posters in Box D as in Box A. How many posters are in Box D?

What happens to the value of each digit when you multiply 231 by 10, 100, or 1,000?

Learn

(a)

$231 \times 10 = 2{,}310$

Box B has _____ posters.

(b)

$231 \times 10 \times 10 = 231 \times 100 = 23{,}100$

Box C has _____ posters.

(c)

$231 \times 10 \times 10 \times 10 = 231 \times 1{,}000 = 231{,}000$

Box D has _____ posters.

Hundred Thousands	Ten Thousands	One Thousands	Hundreds	Tens	Ones
			2	3	1
		2	3	1	0
	2	3	1	0	0
2	3	1	0	0	0

When a number is multiplied by 10, 100, or 1,000, each digit is 10, 100, or 1,000 times as much, respectively. On a place-value chart, each digit moves 1, 2, or 3 places to the left.

What is $231 \times 10{,}000$?
What is $23{,}100 \times 100$?

Do

1 (a) Multiply 24 by 10.

  24 × 10 =

(b) Multiply 24 by 100.

  24 × 100 =

(c) Multiply 24 by 1,000.

  24 × 1,000 =

2

Hundred Thousands	Ten Thousands	One Thousands	Hundreds	Tens	Ones
			3	2	8
		3	2	8	0
	3	2	8	0	0
3	2	8	0	0	0

× 10
× 10
× 10

(a) 328 × 10 =

(b) 328 × 100 =

(c) 328 × 1,000 =

(d) 3,280 × 10 =

(e) 3,280 × 100 =

(f) 32,800 × 10 =

3 Find the values.

(a) 600 × 10   60 × 100   6 × 1,000

(b) 6,000 × 10   600 × 100   60 × 1,000

4 Find the values.

(a) 90 × 10   (b) 415 × 100

(c) 100 × 220   (d) 790 × 1,000

(e) 1,000 × 3,654   (f) 10 × 94,980

(g) 100 × 26,000   (h) 100,000 × 1,000

5 (a) ▓ × 18 = 18,000

(b) 100 × ▓ = 64,000

6

Earth is about 150,000,000 km from the sun. At one point in its orbit, the dwarf planet Eris is about 100 times as far from the sun as Earth is. Write this distance in numerals and in words.

Exercise 2 • page 5

1-2 Multiplying by 10, 100, and 1,000

Lesson 3
Dividing by 10, 100, and 1,000

Think

(a) 231,000 sticker books are divided equally into 10 packages. How many sticker books are in each package?

(b) 231,000 star maps are divided equally into 100 binders. How many star maps are in each binder?

(c) 231,000 science books are divided equally into 1,000 boxes. How many books are in each box?

What happens to the value of each digit when you divide 231,000 by 10, 100, or 1,000?

Learn

(a)

231,00**0** ÷ **10** = 23,100

There are _____ sticker books in each package.

(b)

231,00**0** ÷ **10** ÷ **10** = 231,000 ÷ **100** = 2,310

There are _____ star maps in each binder.

(c)

$231,000 \div 10 \div 10 \div 10 = 231,000 \div 1,000 = 231$

There are _____ science books in each box.

Hundred Thousands	Ten Thousands	One Thousands	Hundreds	Tens	Ones
2	3	1	0	0	0
	2	3	1	0	0
		2	3	1	0
			2	3	1

÷ 100
÷ 1,000
÷ 10
÷ 10
÷ 10

When a number is divided by 10, 100, or 1,000, each digit is $\frac{1}{10}$, $\frac{1}{100}$, or $\frac{1}{1,000}$ times as much, respectively. On a place-value chart, each digit moves 1, 2, or 3 places to the right.

$231,000 \div 1,000 = \frac{231,000}{1,000} = \frac{23,100}{100} = \frac{2,310}{10} = \frac{231}{1} = 231$

$\frac{231,000}{1,000} = 231$

1-3 Dividing by 10, 100, and 1,000

Do

1 (a) Divide 24,000 by 10.

24,000 ÷ 10 = ☐

(b) Divide 24,000 by 100.

24,000 ÷ 100 = ☐

(c) Divide 24,000 by 1,000.

24,000 ÷ 1,000 = ☐

2

Hundred Thousands	Ten Thousands	One Thousands	Hundreds	Tens	Ones
3	2	8	0	0	0
	3	2	8	0	0
		3	2	8	0
			3	2	8

÷ 10
÷ 10
÷ 10

(a) 328,000 ÷ 10 = ☐

(b) 328,000 ÷ 100 = ☐

(c) 328,000 ÷ 1,000 = ☐

(d) 32,800 ÷ 10 = ☐

(e) 32,800 ÷ 100 = ☐

(f) 3,280 ÷ 10 = ☐

3 Find the values.

(a) 600 ÷ 10   6,000 ÷ 100   60,000 ÷ 1,000

(b) 60,000 ÷ 10   600,000 ÷ 100   6,000,000 ÷ 1,000

4 (a) 4,200 ÷ 100 = $\frac{4,200}{100}$ = ⬚

(b) 360,000 ÷ 1,000 = $\frac{360,000}{1,000}$ = ⬚

5 (a) 36,000 ÷ ⬚ = 3,600   (b) 800,000 ÷ ⬚ = 800

6 Find the values.

(a) 560 ÷ 10   (b) 180,000 ÷ 1,000

(c) 10,300 ÷ 100   (d) 79,000,000 ÷ 1,000

7 On a certain day, an asteroid is 1,000 times as far from the sun as Mercury is from the sun. If the asteroid is 57,900,000,000 km from the sun, how far is Mercury from the sun? Write this distance in numerals and in words.

Exercise 3 • page 8

1-3 Dividing by 10, 100, and 1,000

Lesson 4
Multiplying by Tens, Hundreds, and Thousands

Think

A telescope costs $43.

(a) The astronomy club wants to buy 20 telescopes. How much will it cost?

(b) The school wants to buy 200 telescopes. How much will it cost?

(c) The planetarium wants to buy 2,000 telescopes. How much will it cost?

Learn

(a) Multiply 43 by 20.

20 = 2 × 10

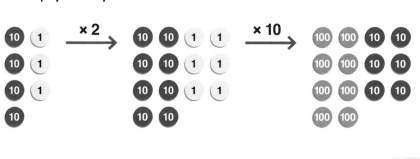

43 × 2 = 86

86 × 10 = 860

43 × 20 = 860

43 × 2 = 80 + 6
         /   \
        40   3

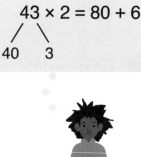

It will cost $_____ to buy 20 telescopes.

(b) Multiply 43 by 200.

200 = 2 × 100

43 × 2 = 86

86 × 100 = 8,600

43 × 200 = 8,600

It will cost $_____ to buy 200 telescopes.

(c) Multiply 43 by 2,000.

2,000 = 2 × 1,000

43 × 2 = 86

86 × 1,000 = 86,000

43 × 2,000 = 86,000

It will cost $_____ to buy 2,000 telescopes.

Do

1 (a) Multiply 27 by 30.

$27 \times 30 = 27 \times 3 \times 10$

$= \boxed{} \times 10$

$= \boxed{}$

(b) Multiply 27 by 300.

$27 \times 300 = 27 \times 3 \times 100$

$= \boxed{} \times 100$

$= \boxed{}$

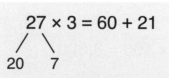

$27 \times 3 = 60 + 21$
 / \
20   7

(c) Multiply 27 by 3,000.

$27 \times 3{,}000 = 27 \times 3 \times 1{,}000$

$= \boxed{} \times 1{,}000$

$= \boxed{}$

2 Multiply 400 by 800.

$400 \times 800 = 400 \times 8 \times 100$

$= \boxed{} \times 100$

$= \boxed{}$

We could also calculate like this:
$400 \times 800 = 4 \times 8 \times \mathbf{100} \times \mathbf{100}$
$ = 32 \times \mathbf{10{,}000}$

3. Multiply 5,000 by 4,000.

   5,000 × 4,000 = 5,000 × 4 × 1,000

   = ⬚ × 1,000

   = ⬚

   We could also calculate like this:
   5,**000** × 4,**000**
   = 5 × 4 × 1,**000** × 1,**000**
   = 20 × 1,**000,000**

4. Multiply.

   (a) 7 × 9   7 × 90   7 × 900   7 × 9,000

   (b) 24 × 6   24 × 60   24 × 600   24 × 6,000

   (c) 360 × 5   360 × 50   360 × 500   360 × 5,000

5. Multiply.

   (a) 70 × 80   (b) 500 × 40   (c) 600 × 800

   (d) 3,000 × 700   (e) 2,000 × 4,000   (f) 9,000 × 6,000

6. The International Space Station can travel a distance of 17,150 miles in one hour.

   (a) How far can the space station travel in 10 hours?

   (b) How far can the space station travel in 100 hours?

Exercise 4 • page 11

Lesson 5
Dividing by Tens, Hundreds, and Thousands

Think

A planetarium committee has a budget of $86,000.

(a) If they spend it all on books for the gift shop that cost $20 each, how many can they buy?

(b) If they spend it all on VR headsets that cost $200 each, how many can they buy?

(c) If they spend it all on special theater chairs that cost $2,000 each, how many can they buy?

Learn

(a) Divide 86,000 by 20.

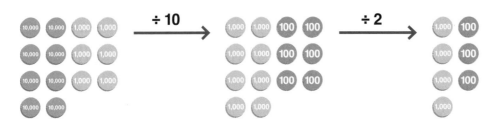

86,000 ÷ 10 = 8,600

8,600 ÷ 2 = 4,300

86,000 ÷ 20 = 4,300

$$\frac{86,000}{20} = \frac{8,600}{2} = \frac{4,300}{1}$$

They could buy _____ books.

(b) Divide 86,000 by 200.

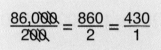

86,000 ÷ 100 = 860

860 ÷ 2 = 430

86,000 ÷ 200 = 430

They could buy _____ VR headsets.

(c) Divide 86,000 by 2,000.

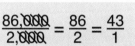

86,000 ÷ 1,000 = 86

86 ÷ 2 = 43

86,000 ÷ 2,000 = 43

They could buy _____ theater chairs.

Do

1 (a) Divide 12,000 by 30.

12,00**0** ÷ **3**0 = 12,000 ÷ 10 ÷ 3

= ☐ ÷ 3

= ☐

(b) Divide 12,000 by 300.

12,0**00** ÷ **3**00 = 12,000 ÷ 100 ÷ 3

= ☐ ÷ 3

= ☐

(c) Divide 12,000 by 3,000.

12,**000** ÷ **3**,**000** = 12,000 ÷ 1,000 ÷ 3

= ☐ ÷ 3

= ☐

2 Divide 400,000 by 800.

400,0**00** ÷ **8**00 = 400,000 ÷ 100 ÷ 8

= ☐ ÷ 8

= ☐

3. Divide 960,000 by 3,000.

   960,000 ÷ 3,000 = 960,000 ÷ 1,000 ÷ 3

   = ▭ ÷ 3

   = ▭

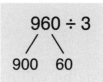

4. Divide.

   (a) 56 ÷ 8   560 ÷ 80   56,000 ÷ 8,000   56,000 ÷ 800

   (b) 40 ÷ 5   400 ÷ 50   40,000 ÷ 50   400,000 ÷ 500

5. Divide.

   (a) 240 ÷ 40   (b) 900 ÷ 90

   (c) 4,800 ÷ 80   (d) 49,000 ÷ 700

   (e) 640,000 ÷ 4,000   (f) 4,500,000 ÷ 9,000

6. A spacecraft traveling to Mars can travel up to 1,600,000 km in 100 hours. How far can it travel in 1 hour?

Exercise 5 • page 14

1-5 Dividing by Tens, Hundreds, and Thousands

Lesson 6
Practice

1 What is the value of each 5 in the number 525,572,505?

2 The approximate distances of Saturn's five farthest moons are given below.

Moon	Approximate Distance from Saturn (km)
Titan	1,221,830
Phoebe	12,952,000
Rhea	527,040
Hyperion	1,481,100
Iapetus	3,561,300

(a) List the moons in order from least to greatest distance.

(b) Write each distance in words.

3 Multiply or divide.

(a) 985 × 100 (b) 1,000 × 785

(c) 85 × 300 (d) 8,000 × 300

(e) 95,000 ÷ 100 (f) 5,600 ÷ 80

(g) 250,000 ÷ 1,000 (h) 3,200,000 ÷ 2,000

Exercise 6 • page 17

Chapter 2

Writing and Evaluating Expressions

Lesson 1
Expressions with Parentheses

Think

The Astronomy Club raised $1,000. They spent $450 on tickets to the space museum and paid $150 for a bus rental. How much money do they have left?

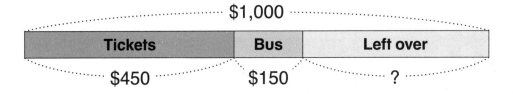

What different math expressions can we write to find the answer?

I wonder if we could show all the steps in one expression.

Learn

Method 1

1,000 − 450 = 550

550 − 150 = 400

"I subtracted the amount spent on tickets first, then the amount spent on the bus rental."

"We could show both subtractions in one expression and then calculate from left to right."

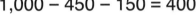

1,000 − 450 − 150 = 400

Method 2

450 + 150 = 600

1,000 − 600 = 400

"I added the cost of the tickets and the bus rental together, and then subtracted that from the total."

1,000 − (450 + 150)

= 1,000 − 600

= 400

"If we use parentheses, we can show this method in one expression. Parentheses indicate which calculation to do first."

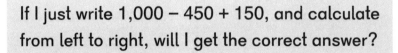

"If I just write 1,000 − 450 + 150, and calculate from left to right, will I get the correct answer?"

They have $_____ left.

Do

1 (a) 100 − 50 + 2

= ▢ + 2

= ▢

(b) 100 − (50 + 2)

= 100 − ▢

= ▢

> Calculate from left to right.
> If there are parentheses, calculate what is in parentheses first.

2 (a) 56 ÷ 2 + 5

= ▢ + 5

= ▢

(b) 56 ÷ (2 + 5)

= 56 ÷ ▢

= ▢

3 Find the values.

(a) 40 − (5 + 5)

(b) 430 − (100 − 20)

(c) 460 + (780 − 250)

④ Find the values.

(a) 9 × (86 − 16)

(b) 800 ÷ (20 + 60)

(c) 1,000 ÷ (48 ÷ 6)

⑤ Rita had a $100 bill. She bought a shirt that cost $25 and a sweater that cost $18. Write an expression with parentheses to find the amount of change she received, and then find the value.

Money she had − (cost of shirt + cost of sweater)

⑥ A chair costs $33 and a chair cushion costs $7. They are sold as a set. Mr. Tomas spent $240 on chair and cushion sets. Write an expression with parentheses to find the number of sets he bought, and then find the value.

Money he spent ÷ (cost of chair + cost of cushion)

Exercise 1 • page 19

2-1 Expressions with Parentheses

Lesson 2
Order of Operations — Part 1

Think

Emma saw a poster with stars on it and thought of a way to find the total number of yellow stars without counting them one by one.

I found the total stars using multiplication and then subtracted 4 groups of 3 red stars.
5 × 5 = 25
4 × 3 = 12
25 − 12 = 13

Write one math expression that shows all the steps in her solution.

Learn

(5 × 5) − (4 × 3)

= 25 − 12

= 13

We learned that we can use parentheses to show which calculation to do first.

We can also write the expression without parentheses if we know that we should multiply 5 × 5 and 4 × 3 first before subtracting.

5 × 5 − 4 × 3

= 25 − 12

= 13

Order of operations
Do multiplication and/or division from left to right, then addition and/or subtraction from left to right.

What other ways can you find? Combine your steps in a single expression.

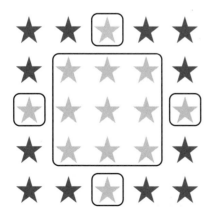

I saw 4 stars on the edges and then 3 groups of 3 stars in the middle.

4 + **3 × 3**
= 4 + 9
= 13

Do

1 Find the values.

(a) 100 + 50 − 2

= ▢ − 2

= ▢

The expressions have only addition and/or subtraction. Calculate from left to right.

(b) 100 − 50 + 2

= ▢ + 2

= ▢

(c) 10 × 5 ÷ 2

= ▢ ÷ 2

= ▢

The expressions have only multiplication and/or division. Calculate from left to right.

(d) 10 ÷ 5 × 2

= ▢ × 2

= ▢

(e) 100 − 50 × 2

= 100 − ▢

= ▢

The expressions have multiplication or division as well as addition and/or subtraction. Multiply or divide first.

2-2 Order of Operations — Part 1

(f) 100 + 50 ÷ 2

= 100 + ▭

= ▭

(g) 56 − 8 × 5 + 12

= 56 − ▭ + 12

= ▭ + 12

= ▭

2 Find the values.

(a) 40 + 3 × 7

(b) 50 + 20 ÷ 4

(c) 3 × 50 + 4 × 30

(d) 60 ÷ 5 − 15 ÷ 5

(e) 96 ÷ 8 − 6 × 2

(f) 22 + 8 ÷ 2 − 2

3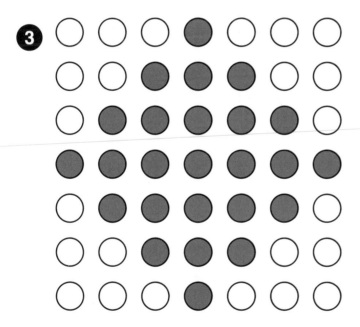

Write a math expression without parentheses that shows all the steps you used in finding the total number of colored dots in this figure, and then show how to find the value.

Exercise 2 • page 22

Lesson 3
Order of Operations — Part 2

Think

Dion is going to make a poster with stickers. He found the perimeter of the poster board by first adding the length and width together, and then multiplying the sum by 2.

50 cm

70 cm

50 + 70 = 120

120 × 2 = 240

The perimeter of the poster board is 240 cm.

How can he show his steps in a single expression?

Learn

(50 + 70) × 2

= 120 × 2

= 240

If I don't use parentheses, and write 50 + 70 × 2, it will be 50 + 140, which is not correct. Use parentheses to show what we should calculate first.

Order of operations
- Calculate the value in parentheses first.
- Multiply and/or divide, from left to right.
- Then add and/or subtract, from left to right.

Do

1 Find the values.

(a) 40 − 25 ÷ 5 × 3 + 12

= 40 − ▢ × 3 + 12

= 40 − ▢ + 12

= ▢ + 12

= ▢

(b) (40 − 25 ÷ 5) × 3 + 12

= (40 − ▢) × 3 + 12

= ▢ × 3 + 12

= ▢ + 12

= ▢

2 Find the values.

Find the value of the expression in parentheses first, using the correct order of operations.

(a) 30 − 6 × 8 ÷ 3

(b) 30 + 10 × 6 − 34 ÷ 2

(c) 100 − 15 + 36 ÷ 4 × 3

(d) 40 ÷ 2 × 4 − 3 × 5

(e) 2 × (36 ÷ 4) × (6 − 2)

(f) (50 ÷ 5) × (4 + 3) − 20

(g) 8 × (10 − 36 ÷ 9) + 2 − 2 × 5 × 5

Exercise 3 • page 25

Lesson 4
Other Ways to Write and Evaluate Expressions

Think

Alex has a sheet with 6 rows of 15 glow-in-the-dark star stickers. He used 2 rows for his poster. Write 2 different expressions, one with parentheses and one without, for finding the number of stars he has left.

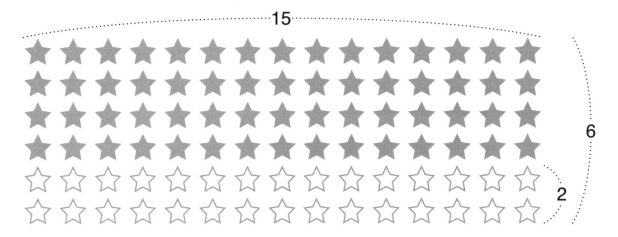

Learn

$(6 - 2) \times 15 = 4 \times 15$

= ▢

$6 \times 15 - 2 \times 15 = 90 - 30$

= ▢

Both expressions have the same value.
$(6 - 2) \times 15 = 6 \times 15 - 2 \times 15$

The parentheses tell us we are multiplying the value of the expression, 6 − 2, by 15. We get the same answer if we multiply each number, 6 and 2, by 15 and then find the difference.

Do

1 Write an expression to show the number that is 3 times as much as the sum of 40 and 6. Then, find the value.

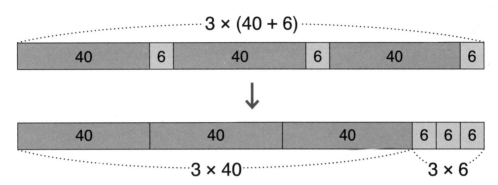

$3 \times (40 + 6) = 3 \times 40 + 3 \times 6$

= ⬚ + ⬚

= ⬚

We can use this idea to mentally calculate 3×46.

2 Write an expression to find the number that is 4 times as much as the difference between 50 and 1. Then, find the value.

$4 \times (50 - 1) = 4 \times 50 - 4 \times 1$

= ⬚ − ⬚

= ⬚

Is this the same as 4×49?

```
    4 9
×     4
```

3 Find the value of 4 × 27 using mental calculation.

4 × (20 + 7)

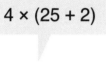
4 × (25 + 2)

4 × (30 − 3)

4 × 27 = ☐

4 (a) 98 × 3 = (100 − 2) × 3 = ☐ × 3 − ☐ × 3

= ☐ − ☐

= ☐

(b) 25 × 49 = 25 × (50 − 1) = 25 × 50 − 25 × 1

= ☐ − ☐

= ☐

(c) 998 × 15 = (1,000 − 2) × 15 = 1,000 × 15 − 2 × 15

= ▭ − ▭

= ▭

(d) 126 × 4 = (101 + 25) × 4 = 101 × 4 + 25 × 4

= ▭ + ▭

= ▭

5 (a) 8 × 82 = 8 × (80 + 2) = ▭

(b) 9 × 99 = 9 × (100 − 1) = ▭

(c) 32 × 49 = 32 × (50 − ▭) = ▭

(d) 4,998 × 6 = (5,000 − ▭) × 6 = ▭

6 What sign, >, <, or =, goes in each ◯?

(a) (14 + 27) × 6 ◯ 8 × (27 + 14)

(b) 8 × (48 − 7) ◯ 48 × 8 − 9 × 8

(c) 9 × 38 − 9 × 9 ◯ 9 × (38 − 9)

Exercise 4 • page 28

Lesson 5
Word Problems — Part 1

Think

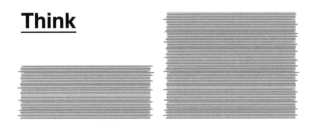

Ani, Eliza, and Tiara have $490 altogether. Eliza has 3 times as much money as Ani. Tiara has $40 more than Ani. How much money does Tiara have?

Learn

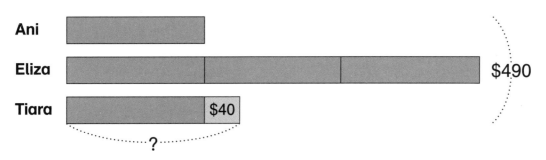

5 units ⟶ 490 − 40
1 unit ⟶ (490 − 40) ÷ 5 = 450 ÷ 5 = 90
90 + 40 = 130

Tiara has $_____.

Do

1 George bought 2 identical couches and 3 identical chairs. A couch cost twice as much as a chair. He spent $270 more on the couches than on the chairs. How much did he spend altogether?

Couches

Chairs

?

$270

2 The total weight of 3 suitcases is 75 lb. Suitcase B weighs 15 lb more than Suitcase A. Suitcase C weighs 6 lb less than Suitcase A. How much does Suitcase C weigh?

Suitcase A

Suitcase B — 75 lb

15 lb

Suitcase C

? 6 lb

Emma and Mei solved the problem in different ways. Complete their steps.

I used Suitcase A as the unit.
3 units ⟶ 75 − 15 + 6 = ?
1 unit ⟶ ...
Suitcase C: ...

I used Suitcase C as the unit, since that is what we need to find.
3 units ⟶ 75 − 15 − (2 × 6) = ?
1 unit ⟶ ...

3) A library spent $6,000 on a printer and 5 laptops. Each laptop cost $420 more than the printer. How much does the printer cost?

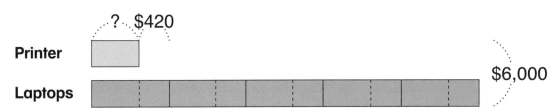

4) The museum sells picture books about Mars for $18 each. They have 5 boxes with 4 books in each box. How much money will they receive if they sell all of the books?

5) A 1,500 cm long rope was cut into 3 smaller ropes, A, B, and C. Rope B is 300 cm longer than Rope A. Rope C is 300 cm longer than Rope B. How long is Rope C?

Exercise 5 • page 31

Lesson 6
Word Problems — Part 2

Think

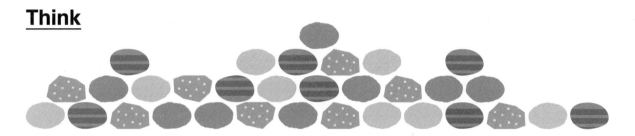

At first, Liam had 20 more polished rocks than Carlos. Then, Carlos gave 30 polished rocks to Liam. Now, Liam has twice as many polished rocks as Carlos. How many polished rocks did Liam have at first?

Learn

Before:

Liam [_____ 20]

Carlos [_____]

After: 1 unit · · · 1 unit

Liam [_____ | 30 | 20 | 30]

Carlos [_____ | 30]

1 unit ⟶ 30 + 20 + 30 = 80

2 units ⟶ 80 × 2 = 160

160 − 30 = 130

Liam had _____ polished rocks at first.

Do

1 A container with wheat flour weighs 90 kg and an identical container with rye flour weighs 50 kg. The wheat flour weighs 3 times as much as the rye flour. How much does the container weigh?

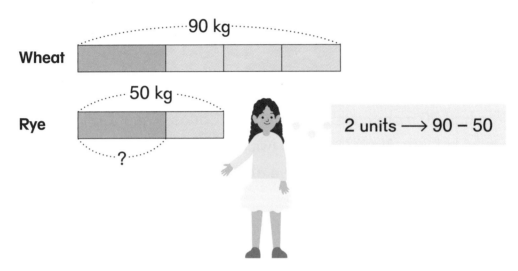

2 units ⟶ 90 − 50

2 Jasmine had twice as much money as her sister. After giving her sister $45, they each had the same amount of money. How much money did Jasmine have at first?

Before:

Jasmine

Sister

After:

Jasmine $45

Sister $45

1 unit ⟶ 2 × 45

3

A bouquet of 2 gerberas and 2 sunflowers costs $18. A bouquet of 2 gerberas and 4 sunflowers costs $24. How much does one sunflower cost? How much does one gerbera cost?

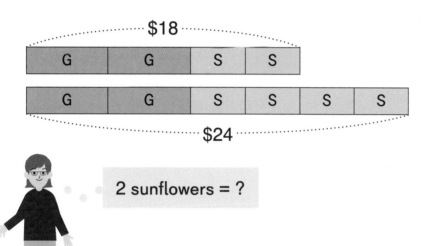

2 sunflowers = ?

4 Emma and Alex have 980 coins altogether. Emma has 4 times as many coins as Alex. How many coins would she have to give Alex for them to have the same number of coins?

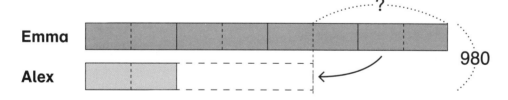

5 A jug and a bottle had a combined total of 950 mL of water. After 50 mL of water was poured from the bottle to the jug, the jug had 4 times as much water as the bottle. How much water was in each container at first?

Exercise 6 • page 36

Lesson 7
Practice

1 What sign, >, <, or =, goes in each ◯?

(a) (7 × 1,000,000) + (8 × 10,000) ◯ 760,000,000

(b) (50 + 30) × 9 ◯ 9 × (100 − 20)

(c) 7 × (100 − 8) ◯ 98 × 7

(d) 500,000 × 400 − 500,000 × 40 ◯ 500,000 × 320

2 Find the values.

(a) 3,000 − (575 + 128)

(b) (600 + 1,400) ÷ 40

(c) 235 + 37 × 6

(d) 5,000 − 800 ÷ 4

(e) (1,200 − 400) × (45 − 5 × 3)

(f) (40 ÷ 5 − 6) + 90 ÷ 9 × 7

(g) 2,000 × (25 + 15 × 2)

(h) (60,000 + 20,000) × 90 ÷ (1,500 − 700)

③ The friends below wrote some of their steps when they calculated 5 + 90 ÷ 5 × 2 − 6. Whose answer is correct? What did the others do wrong?

5 + 90 ÷ 5 × 2 − 6 = 95 ÷ 5 × 2 − 6 = 19 × 2 − 6 = 32	5 + 90 ÷ 5 × 2 − 6 = 5 + 90 ÷ 10 − 6 = 5 + 9 − 6 = 8	5 + 90 ÷ 5 × 2 − 6 = 5 + 18 × 2 − 6 = 5 + 36 − 6 = 35

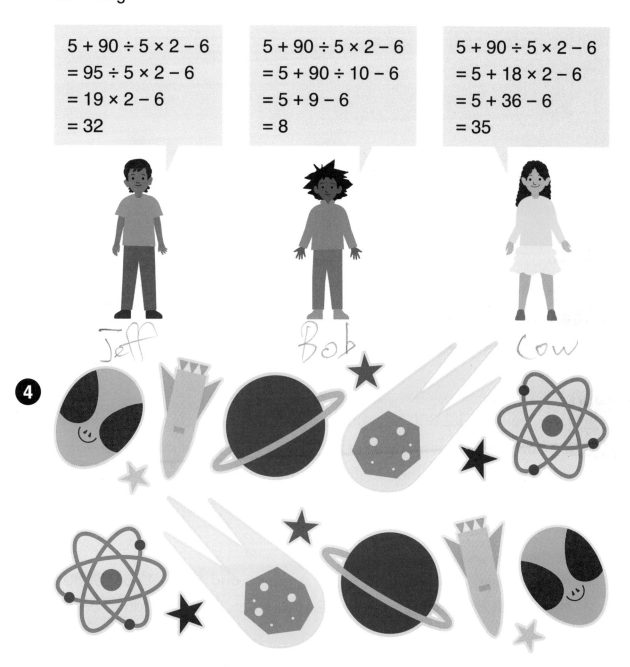

④ Abigail got 50 glow-in-the-dark stickers and 2 packs of 20 neon stickers from the planetarium gift shop. She gave 15 stickers to each of her 2 brothers and 5 stickers to each of her 5 friends. How many stickers does she have left?

5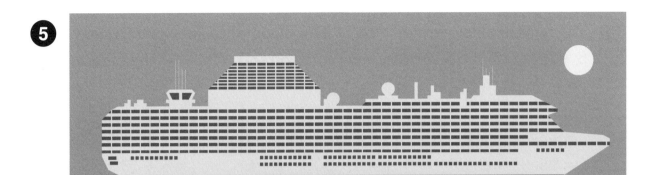

There are 3 times as many girls as boys on a cruise ship. There are twice as many adults as girls. There are 4,500 people altogether on the cruise ship. How many adults are there?

6 Kaylee and Heather had the same amount of money at first. After Kaylee spent $90, Heather had 4 times as much money as Kaylee. How much money did they have altogether at first?

7 Ryan collected 2,650 baseball cards, football cards, and hockey cards. He has 3 times as many baseball cards as football cards. He has 150 more hockey cards than football cards. How many baseball cards does he have?

8 A laptop and 3 tablets cost $810. A laptop and 1 tablet cost $630. How much does one laptop cost? How much does one tablet cost?

9 Box A and Box B contained a total of 18,000 nails. After 2,500 nails were transferred from Box A to Box B, Box A had 3,000 more nails than Box B. How many nails did Box A have at first?

Exercise 7 • page 41

Chapter 3

Multiplication and Division

Lesson 1
Multiplying by a 2-digit Number — Part 1

(1)

Think

A saddle costs $86.

(a) Estimate the cost of 25 saddles.

(b) Find the actual cost of 25 saddles.

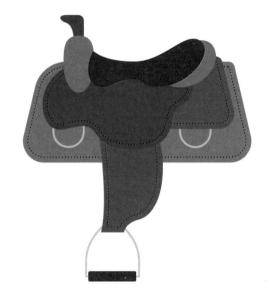

Learn

86 × 25

(a) We can estimate the product in different ways.

86 × 25	86 × 25	86 × 25
↓ ↓	↓ ↓	↓ ↓
90 × 30 = 2,700	90 × 20 = 1,800	80 × 25 = 2,000

Round to close numbers that are easy to calculate. Sometimes we can get a better estimate by rounding one number down and the other number up. If the calculation is easy, we might round only one number.

86 × 25 ≈ ▢

≈ stands for **is approximately equal to** or **is about**.

(b) 86 × 25 = 86 × 20 + 86 × 5

First, multiply 86 by 5.

```
    86
×   25
   430
```

Then, multiply 86 by 20.

```
    86
×   25
   430
 1,720
```

To multiply 86 by 2 tens, we can first write a **0** in the ones place and then multiply 86 by 2.

Then, add the products.

```
    86
×   25
   430  ← 86 × 5
 1,720  ← 86 × 20
 2,150  ← 86 × 25
```

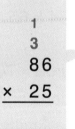

We can write the regrouped digits above the number we will multiply next, to help remember to add them.

```
   1
   3
    86
×   25
```

Which estimate was closest to the actual answer? Is the answer reasonable?

The 25 saddles cost $_____.

Do

1 Estimate the products. Then find the actual products.

(a) 53 × 32 ≈ ▢

```
    5 3
  × 3 2
  ─────
    ▢     ← 53 × 2
   ▢      ← 53 × 30
  ─────
   ▢
```

To multiply 53 by 3 tens, we can first write 0 in the ones place, then find 53 × 3.

```
    5 3
  × 3 2
  ─────
   1 0 6
    ▢ 0
       ↑
     53 × 3
```

(b) 209 × 38 ≈ ▢

```
    2 0 9
  ×   3 8
  ───────
    ▢     ← 209 × 8
   ▢      ← 209 × 30
  ───────
   ▢
```

209 × 38
 ↓   ↓
200 × 40

(c) 584 × 65 ≈ ▢

600 × 70 will be much greater than 584 × 65, so to get a better estimate we could use 600 × 60.

```
    5 8 4
  ×   6 5
  ───────
    ▢
   ▢
  ───────
   ▢
```

50   3-1 Multiplying by a 2-digit Number — Part 1

2 Estimate the products and then find the actual products.

(a) 48 × 72

(b) 97 × 68

(c) 57 × 45

(d) 94 × 36

(e) 307 × 44

(f) 760 × 89

(g) 235 × 84

(h) 706 × 14

(i) 387 × 58

(j) 23 × 15 × 37

3 A ton of alfalfa costs $350. Mei, Emma, and Alex estimated the cost of 25 tons of alfalfa in different ways.

Mei
350 × 25 ≈ 400 × 30
         = $12,000

Emma
350 × 25 ≈ 400 × 20
         = $8,000

Alex
350 × 25 ≈ 300 × 30
         = $9,000

Which estimate do you think will be closest to the actual cost?

(a) Find the actual cost of 25 tons of alfalfa.

(b) Explain why Emma's and Alex's estimates were closer to the actual cost than Mei's estimate.

Exercise 1 • page 45

3-1 Multiplying by a 2-digit Number — Part 1

Lesson 2
Multiplying by a 2-digit Number — Part 2

Think

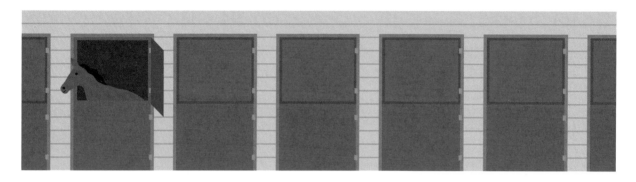

A stable boards 23 horses. The stable receives $8,416 per year to board each horse. How much money will the stable receive to board the horses for a year?

(a) Estimate the amount.

(b) Find the actual amount.

Learn

8,416 × 23

(a)

8,416 × 23
↓   ↓
8,000 × 20 = 160,000

8,416 × 23
↓   ↓
8,000 × 25 = 200,000

Which estimate do you think will be closer to the actual product?

8,416 × 23 ≈ ☐

(b) 8,416 × 23 = (8,416 × 20) + (8,416 × 3)

First, multiply 8,416 by 3.

```
    8,416
×      23
   25,248
```

Then, multiply 8,416 by 20.

```
    8,416
×      23
   25,248
  168,320
```

To multiply by 2 tens, we can first write a **0** in the ones place and then multiply by 2.

We can write the regrouped digits on top.

```
       1
     1 1
   8,416
×     23
```

Then, add the products.

```
    8,416
×      23
   25,248   ← 8,416 × 3
  168,320   ← 8,416 × 20
  193,568   ← 8,416 × 23
```

Compare the actual product to your estimate. Is the answer reasonable?

The stable will receive $_____.

3-2 Multiplying by a 2-digit Number — Part 2

Do

1 Estimate the products. Then find the actual products.

(a) 2,384 × 32 ≈ ☐

```
    2,3 8 4
  ×     3 2
  ─────────
   ☐       ← 2,384 × 2
   ☐       ← 2,384 × 30
  ─────────
   ☐       ← 2,384 × 32
```

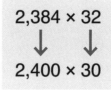

2,384 × 32
↓   ↓
2,400 × 30

(b) 5,302 × 74 ≈ ☐

```
    5,3 0 2
  ×     7 4
  ─────────
   ☐       ← 5,302 × 4
   ☐       ← 5,302 × 70
  ─────────
   ☐       ← 5,302 × 74
```

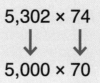

5,302 × 74
↓   ↓
5,000 × 70

(c) 38,071 × 49 ≈ ☐

```
    3 8,0 7 1
  ×       4 9
  ───────────
   ☐
   ☐
  ───────────
   ☐
```

2 Estimate the products and then find the actual products.

(a) 5,438 × 50   (b) 70 × 6,908   (c) 83 × 5,600

(d) 3,470 × 25   (e) 6,804 × 68   (f) 7,566 × 98

3

Horseshoes come in boxes of 20 pairs. The farmer bought 328 boxes.

(a) How many pairs of horseshoes did he buy?

(b) If horseshoes cost $6 per pair, how much did the farmer spend altogether?

4 It costs $112 per month to shoe a show horse, and $139 to shoe a draft horse.

(a) Estimate the monthly cost to shoe 35 show horses and 12 draft horses.

(b) Find the actual monthly cost.

Exercise 2 • page 49

Lesson 3
Practice A

1 Estimate the values and then find the actual values.

(a) 36 × 63

(b) 75 × 208

(c) 5,019 × 32

(d) 6,819 × 48

(e) 17 × 2,175

(f) 4,087 × 46

(g) 99 × 9,999

(h) 5,630 × 78

(i) 7,092 × 48

(j) 60,812 × 82

(k) 16 × 125 × 8

(l) 32 × 7 × 128

2 Mr. Johnson made a down payment of $1,500 for a car. He will make monthly car payments of $325 for 48 months. Estimate and then find the total cost of the car.

3 A horse blanket costs $89 and a grooming kit costs $15. The owner wants to buy 25 of each item. Estimate and then find the actual cost.

4 The stable wants to be able to transport 12 horses at a time. A 4-horse trailer costs $48,900. A 2-horse trailer costs $16,400.

(a) Would it cost less to buy 4-horse trailers only or to buy 2-horse trailers only?

(b) What would be the difference in cost?

Exercise 3 • page 52

Lesson 4
Dividing by a Multiple of Ten

Think

There are 90 lb of oats. Emma wants to put 20 lb in each bag. How many bags will she use? How many pounds of oats will be left over?

Learn

Divide 90 into groups of 20.

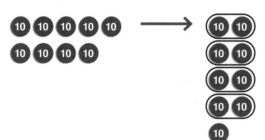

Estimate the quotient first.
90 ÷ 20 ≈ 80 ÷ 20 = 4

```
      4
20)90
    80   ← 20 × 4
    10   ← 90 − 80
```

20 × 4 is close to but not greater than 90.

The **divisor** is the number we are dividing by. The **dividend** is the number we are dividing.

Why does the remainder have to be less than the divisor?

She will use _____ bags. _____ lb of oats will be left over.

Check: 4 × 20 + 10 = ▢

Do

1 Divide 80 by 30.

$80 \div 30 \approx 90 \div 30 = 3$
$3 \times 30 = 90$ so the estimated quotient is too big. Try 2.

```
      2
30)80
    60   ← 30 × 2
    20   ← 80 − 60
```

80 ÷ 30 is ▢ with remainder of ▢.

We can write this as 80 ÷ 30 is 2 R 20.

2 Divide 150 by 40.

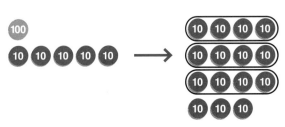

$160 \div 40 = 4$
40×4 is greater than 150. Try 3.

```
40)150
    ▢    ← 40 × ▢
    ▢    ← 150 − 120
```

How many groups of 4 tens can we make with 15 tens?

3-4 Dividing by a Multiple of Ten

3 Divide 89 by 20.

20)89

80 ÷ 20 = 4
20 × 4 is less than 89.

4 Divide 625 by 70.

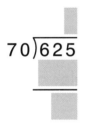

70)625

630 ÷ 70 = 9
70 × 9 is greater than 625. Try…

5 Divide.

(a) 70 ÷ 20

(b) 90 ÷ 40

(c) 79 ÷ 30

(d) 97 ÷ 20

(e) 370 ÷ 50

(f) 420 ÷ 80

(g) 587 ÷ 70

(h) 439 ÷ 60

6 279 people are going to the horse show on buses. 50 people can go on one bus. How many buses will they need?

Exercise 4 • page 56

Lesson 5
Divide a 2-digit Number by a 2-digit Number

Think

The friends are preparing snack bags for the horses. Sofia has 86 apples. She wants to put 21 apples in each bag. How many bags of apples can she make? How many apples will be left over?

Learn

Divide 86 into groups of 21.

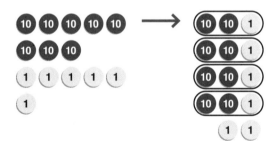

How many groups of 21 can we make with 86?

```
      4
21)86
    84  ← 21 × 4
     2  ← 86 − 84
```

$86 ÷ 21 ≈ 80 ÷ 20 = 4$
Try 4. 21 × 4 is close to but not greater than 86.

Check: 4 × 21 + 2 = _____

She can make _____ bags of apples. _____ apples will be left over.

Do

1 Divide 98 by 31.

$$\begin{array}{r} 3 \\ 31\overline{)98} \\ \square \\ \hline \square \end{array}$$

← 31 × 3

← 98 − 93

Check: 31 × 3 + 5 = ?

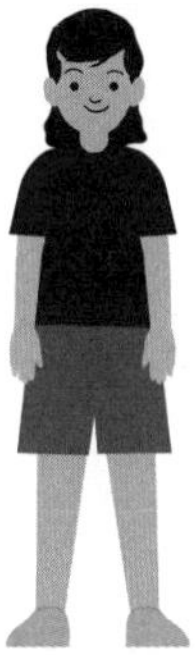

90 ÷ 30 = 3

31 × 3 is close to but less than 98.

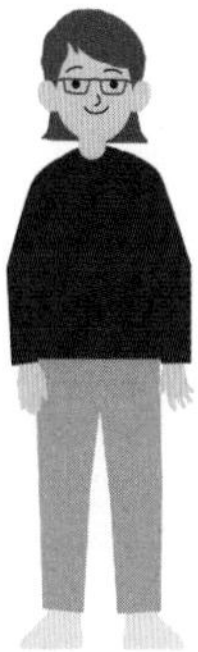

2 Divide 81 by 16.

$$\begin{array}{r} \square \\ 16\overline{)81} \\ \square \\ \hline \square \end{array}$$

80 ÷ 20 = 4

$$\begin{array}{r} 4 \\ 16\overline{)81} \\ \underline{64} \\ 17 \end{array} \longrightarrow \begin{array}{r} 5 \\ 16\overline{)81} \end{array}$$

The remainder is greater than 16 so the estimated quotient is too small. Try 5.

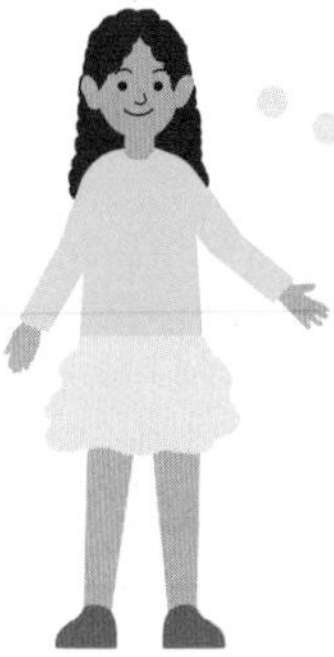

3 Divide.

(a) 78 ÷ 26

(b) 84 ÷ 12

(c) 85 ÷ 17

(d) 89 ÷ 43

(e) 98 ÷ 24

(f) 95 ÷ 13

(g) 83 ÷ 37

(h) 85 ÷ 27

(i) 95 ÷ 16

(j) 89 ÷ 24

Check your answers.

4 Carlton Horse Barn stables 91 horses. Their daily exercise sessions are divided equally among the 13 trainers. How many horses does each trainer exercise each day?

5 Alex has 86 hay cubes. He wants to put 15 hay cubes in each bag.

(a) How many bags of hay cubes can he make?

(b) How many more hay cubes does he need to make another bag?

Exercise 5 • page 59

Lesson 6
Divide a 3-digit Number by a 2-digit Number — Part 1

Think

Dion has $154 to buy horse brushes. Each brush costs $31. How many horse brushes can he buy? How much money will he have left?

Learn

Divide 154 into groups of 31.

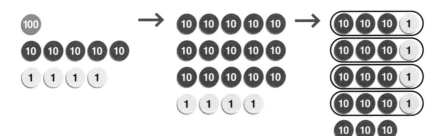

How many groups of 31 can we make with 154?

$154 \div 31 \approx 150 \div 30 = 5$

```
     5
31)154
   155
```

The estimated quotient is too large. Try 4.

```
      4
31)154
   124   ← 31 × 4
    30   ← 154 − 124
```

Check: 31 × 4 + 30 = _____

He can buy _____ brushes. He will have $_____ left.

Do

1 Divide 321 by 38.

$320 \div 40 = 8$

38×8 is close to but less than 321.

```
      8
38)321
```
← 38 × 8
← 321 − 304

Check: 38 × 8 + 17 = ?

2 Divide 285 by 43.

$280 \div 40 = 7$

```
      7
43)285
   301
```

The estimated quotient is too large. Try 6.

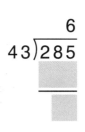

3 Divide 189 by 26.

$180 \div 30 = 6$

```
      6
26)189
   156
    33
```

Since 33 is greater than the divisor, the estimated quotient is too small. Try 7.

4) Divide 183 by 24.

24)183

180 ÷ 20 = 9
The estimated quotient is too large.
If we try 8 it is still too large. Try 7.

```
     9         8         7
24)183    24)183    24)183
   216       192       168
```

5) Divide.

(a) 128 ÷ 25

(b) 147 ÷ 21

(c) 379 ÷ 53

(d) 237 ÷ 39

(e) 163 ÷ 18

(f) 358 ÷ 72

(g) 640 ÷ 92

(h) 308 ÷ 31

6) Mei has 15 boxes that each contain 21 sugar cubes. She wants to put them equally into 45 bags. How many sugar cubes should she put in each bag?

Exercise 6 • page 61

Lesson 7
Divide a 3-digit Number by a 2-digit Number — Part 2

Think

The stable is having an awards luncheon for the riders. The cost of the luncheon is $576. The cost will be shared equally by 16 donors. How much will each donor pay?

Learn

Divide $576 by 16.

Regroup 5 hundreds as 50 tens, then divide 57 tens by 16.

```
      3
16)576
   48      ← 16 × 3 tens = 48 tens
    9      ← 57 tens − 48 tens
```

60 tens ÷ 20 = 3 tens. I will try 3 in the tens place first.

We are dividing the tens, so begin writing the quotient in the tens place.

Regroup 9 tens as 90 ones, then divide 96 ones by 16.

```
     36
16)576
   48
    96
    96     ← 16 × 6 ones
     0     ← 96 ones − 96 ones
```

Since 100 ÷ 20 = 5, I tried 5 in the ones place. It was too small, so then I tried 6.

Check: 16 × 36 = _____

Each donor will pay $_____.

Do

① Divide 284 by 12.

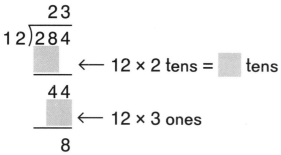

I know 12 × 2 = 24. I will write 2 in the tens place.

⟵ 12 × 2 tens = ▢ tens

⟵ 12 × 3 ones

Check: 12 × ▢ + ▢ = ▢

② Divide 643 by 23.

60 tens ÷ 20 = 3 tens
23 × 3 tens = 69 tens, which is too large. Try 2 in the tens place.

⟵ 23 × 2 tens = ▢ tens

⟵ 23 × ▢ ones

③ Divide 725 by 24.

What should I write in the ones place in the quotient?

4. Determine whether each quotient will be a 1-digit number or a 2-digit number, then divide.

   (a) 464 ÷ 57

   (b) 926 ÷ 43

   Are the first two digits in the dividend greater than or less than the divisor?

   $57\overline{)464}$ | $43\overline{)926}$

   (c) 330 ÷ 27

   (d) 705 ÷ 47

5. Explain the mistakes below. Then do each calculation correctly.

   (a)
   ```
        33
   27)919
       81
      109
       81
       28
   ```

   (b)
   ```
        61
   12)845
       72
       12
       12
        0
   ```

6. Divide.

   Check your answers.

   (a) 546 ÷ 21

   (b) 687 ÷ 36

   (c) 726 ÷ 24

   (d) 900 ÷ 18

7. 17 people are sharing the cost of a meal at a restaurant equally. The bill, including tip, is $544. How much money should each person pay?

Exercise 7 • page 63

Lesson 8
Divide a 4-digit Number by a 2-digit Number

Think

Sofia and Emma are helping the stable owner put 1,344 bales of straw equally in 12 barns. How many bales of straw should they put in each barn?

Learn

Divide 1,344 into 12 equal groups.

13 hundreds ÷ 12 ≈ 12 hundreds ÷ 12
= 1 hundred
The answer will be a little more than 100.

In what place should we begin writing the quotient?

12)‾1,344

Regroup 1 thousand as 10 hundreds and divide 13 hundreds by 12.

```
        1
    _____
12 ) 1,3 4 4
     12         ← 12 × 1 hundred = 12 hundreds
      1         ← 13 hundreds − 12 hundreds
```
↓

Regroup 1 hundred as 10 tens and divide 14 tens by 12.

```
        1 1
    _____
12 ) 1,3 4 4
     12
      1 4
      1 2       ← 12 × 1 ten
        2       ← 14 tens − 12 tens
```
↓

Regroup 2 tens as 20 ones and divide the 24 ones by 12.

```
        1 1 2
    _____
12 ) 1,3 4 4
     12
      1 4
      1 2
        2 4
        2 4    ← 12 × 2 ones
          0    ← 24 ones − 24 ones
```

Check: 12 × 112 = _____

They should put _____ bales of straw in each barn.

3-8 Divide a 4-digit Number by a 2-digit Number

Do

1 Divide 4,362 by 18.

```
        2 4 2
    ┌─────────
18 )  4,3 6 2
      ▢ ▢       ← 18 × 2 hundreds = ▢ hundreds
      ─────
        7 6
        ▢ ▢     ← 18 × 4 tens = ▢ tens
        ─────
          4 2
          ▢ ▢   ← 18 × 2 ones
          ─────
            6
```

Divide 43 hundreds by 18 and begin writing the quotient in the hundreds place.

2 Divide 9,280 by 35.

```
         ▢
       2
    ┌─────────
35 ) 9,2 8 0
     7 0          ← 35 × 2 hundreds = 70 hundreds
     ─────
       2 2 8
       ▢ ▢ ▢      ← 35 × ▢ tens = ▢ tens
       ───────
         ▢ ▢
         ▢ ▢      ← 35 × ▢ ones
         ─────
           ▢
```

3 Divide 8,524 by 28.

```
        3              3 ▢
    ┌─────────      ┌─────────
28 ) 8,5 2 4   →   28 ) 8,5 2 4
     8 4               8 4
     ─────             ─────
       1 2               1 2 4
                         ▢ ▢ ▢
                         ───────
                           ▢
```

We cannot divide 12 tens into 28 groups of tens. What should we write in the tens place in the quotient?

4 Divide 4,528 by 52.

$$52 \overline{)4{,}528}$$
$$416$$

We cannot divide 45 hundreds into 52 groups of hundreds. In what place should we begin writing the quotient?

$$52 \overline{)4{,}528}$$

5 Tell whether each quotient will be a 2-digit number or a 3-digit number. Then divide.

(a) 7,025 ÷ 21

(b) 3,470 ÷ 45

(c) 1,008 ÷ 25

(d) 8,060 ÷ 62

Are the first two digits in the dividend greater than or less than the divisor?

$$21 \overline{)7{,}025} \quad | \quad 45 \overline{)3{,}470}$$

6 Divide.

(a) 8,390 ÷ 34

(b) 8,875 ÷ 29

(c) 4,796 ÷ 57

(d) 4,340 ÷ 62

7 Dion wants to save $1,000. How many weeks will it take him to save that much if he saves $35 each week?

Exercise 8 • page 66

Lesson 9
Practice B

1 Estimate the quotient and then find the actual quotient.

(a) 92 ÷ 50   (b) 78 ÷ 30   (c) 98 ÷ 40

(d) 98 ÷ 32   (e) 62 ÷ 16   (f) 142 ÷ 24

(g) 435 ÷ 70   (h) 650 ÷ 80   (i) 786 ÷ 30

(j) 362 ÷ 58   (k) 1,462 ÷ 47   (l) 2,205 ÷ 72

(m) 8,350 ÷ 18   (n) 6,204 ÷ 29   (o) 95,987 ÷ 96

2

It costs $2,730 for 35 riding helmets. Estimate and then find the cost for 1 helmet.

3 A flour mill produced 1,000 lb of flour. The workers put 300 lb of flour into 25-lb bags and the rest into 50-lb bags. How many of each kind of bag will they have?

4 A school paid $2,212 for 4 projectors and 8 tablets. A tablet cost three times as much as a projector. How much did each projector and each tablet cost?

Exercise 9 • page 69

Chapter 4

Addition and Subtraction of Fractions

Lesson 1
Fractions and Division

Think

Dion has 38 lb of flour. The baker asks him to put the flour equally into 3 containers. How many pounds of flour should he put in each container?

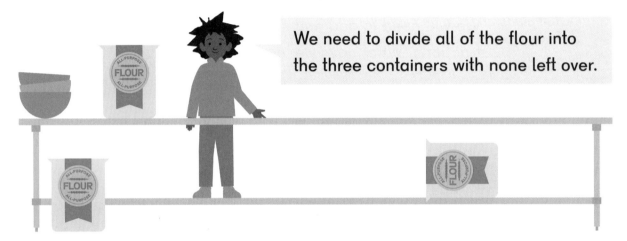

We need to divide all of the flour into the three containers with none left over.

Learn

38 ÷ 3

First, divide 3 tens by 3.

3 tens ÷ 3 is 1 ten.

$$3\overline{\smash{)}38} \\ \underline{3} \\ 8$$

76   4-1 Fractions and Division

Then, divide 8 ones by 3.

8 ones ÷ 3 is 2 ones with 2 ones left over.

```
      1 2
3 ) 3 8
    3
    ─
      8
      6
      ─
      2
```

Then, divide the 2 ones by 3.

2 ones ÷ 3 = $\frac{2}{3}$

We can divide the remainder and express the result as a fraction.

$2 \div 3 = \frac{2}{3}$

$12 + \frac{2}{3} = 12\frac{2}{3}$

$38 \div 3 = 12\frac{2}{3}$

$38 \div 3 = \frac{38}{3}$. Does $12\frac{2}{3} = \frac{38}{3}$?

There will be _____ lb of flour in each container.

4-1 Fractions and Division

Do

1 (a) Express 4 ÷ 6 as a fraction.

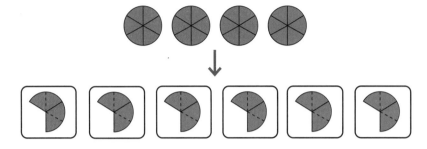

$4 \div 6 = \frac{4}{6} = \frac{\square}{3}$

(b) Express 6 ÷ 4 as a mixed number.

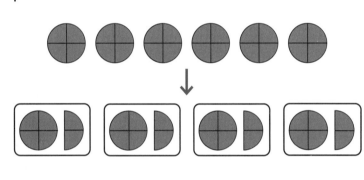

$6 \div 4 = \frac{6}{4} = \square\frac{\square}{4} = \square\frac{\square}{2}$

2 Express 10 ÷ 4 as a mixed number.

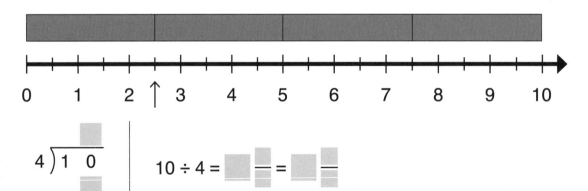

$10 \div 4 = \square\frac{\square}{\square} = \square\frac{\square}{\square}$

3. Express $\frac{27}{8}$ as a mixed number.

```
    ☐
  ┌─────
8 │ 2 7
    2 4
  ─────
```

$\frac{27}{8} = 27 \div 8 = \boxed{}\frac{\boxed{}}{}$

$\frac{27}{8} = \frac{24}{8} + \frac{3}{8}$

4. Express $\frac{100}{6}$ as a mixed number.

```
    ☐ ☐
  ┌─────
6 │ 1 0 0
```

Check: $16\frac{2}{3} = \frac{50}{3} = \frac{100}{6}$

$\frac{100}{6} = 100 \div 6 = \boxed{}\frac{\boxed{}}{6} = \boxed{}\frac{\boxed{}}{3}$

5. Divide 280 by 12. Express the answer as a mixed number.

$\overset{\div 4}{\underset{\div 4}{\frac{280}{12} = \frac{70}{3}}}$

```
    ☐ ☐
  ┌─────
3 │ 7 0
```

If we can simplify first, we can make the division easier.
$280 \div 12 = 70 \div 3$

$280 \div 12 = \boxed{}\frac{\boxed{}}{3}$

4-1 Fractions and Division

6 Find the value of 2,200 ÷ 80. Express the answer as a mixed number in simplest form.

$\frac{2{,}200}{80} = \frac{220}{8} = 220 \div 8 = ?$

I can use mental math if I simplify even more.

$\frac{220}{8} = \frac{110}{4} = \frac{55}{2} = 55 \div 2 = ?$

2,200 ÷ 80 =

7 Divide. Express each answer as a fraction or mixed number in simplest form.

(a) 5 ÷ 7

(b) 9 ÷ 6

(c) 56 ÷ 3

(d) 82 ÷ 12

(e) 120 ÷ 14

(f) 338 ÷ 16

8 Express each fraction as a mixed number in simplest form.

(a) $\frac{8}{5}$

(b) $\frac{13}{4}$

(c) $\frac{88}{12}$

(d) $\frac{40}{15}$

9 Emma cut 53 ft of ribbon into 3 equal length pieces. What is the length of each piece in feet?

Exercise 1 • page 73

Lesson 2
Adding Unlike Fractions

Think

A cornbread recipe calls for $\frac{1}{3}$ cup of corn flour and $\frac{1}{2}$ cup of wheat flour. How many cups of flour does the recipe call for?

Learn

$\frac{1}{3} + \frac{1}{2}$

6 is a common multiple of 3 and 2.
3, **6**, …
2, 4, **6**, …

$\frac{1 \times 2}{3 \times 2} = \frac{2}{6}$ $\frac{1 \times 3}{2 \times 3} = \frac{3}{6}$

We cannot add different-sized units. We need to find equivalent fractions with the same denominators.

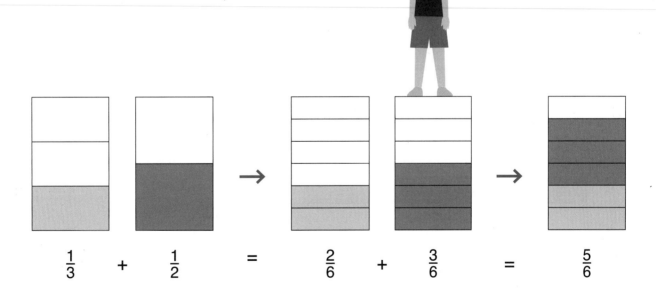

$\frac{1}{3} + \frac{1}{2} = \frac{2}{6} + \frac{3}{6} = \frac{5}{6}$

The recipe calls for _____ cups of flour.

Do

1 Add $\frac{5}{6}$ and $\frac{1}{3}$.

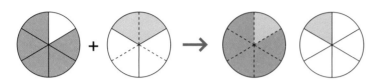

$\frac{5}{6} + \frac{1}{3} = \frac{5}{6} + \frac{\square}{6} = \frac{\square}{6} = 1\frac{\square}{\square}$

2 Add $\frac{2}{3}$ and $\frac{1}{4}$.

3, 6, 9, **12**...
4, 8, **12**...
12 is a common multiple of both denominators.

 $\frac{2 \times 4}{3 \times 4} = \frac{\square}{12}$

 $\frac{1 \times 3}{4 \times 3} = \frac{\square}{12}$

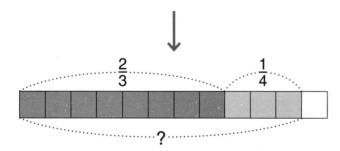

$\frac{2}{3} + \frac{1}{4} = \frac{\square}{12} + \frac{\square}{12} = \frac{\square}{\square}$

3 Add $\frac{1}{2}$ and $\frac{3}{5}$.

$\frac{1 \times 5}{2 \times 5} = \frac{\square}{10}$

$\frac{3 \times 2}{5 \times 2} = \frac{\square}{10}$

$+ \frac{\square}{10}$

> I can multiply the denominators to find a common multiple.
> 2 × 5 = 10
> 2 and 5 are both factors of 10.

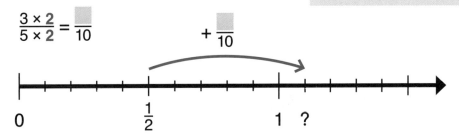

$\frac{1}{2} + \frac{3}{5} = \frac{\square}{10} + \frac{\square}{10} = \square \frac{\square}{\square}$

4 Add $\frac{1}{4}$ and $\frac{1}{6}$. Express the answer in simplest form.

> Will the answer be greater than or less than $\frac{1}{2}$?

(a) $\frac{1}{4} + \frac{1}{6} = \frac{\square}{\square} + \frac{\square}{\square} = \frac{\square}{\square}$

(b) $\frac{1}{4} + \frac{1}{6} = \frac{\square}{\square} + \frac{\square}{\square} = \frac{\square}{\square} = \frac{\square}{\square}$

> I used the least common multiple. 4, 8, **12**... 6, **12**...
> $\frac{1 \times 3}{4 \times 3} = \frac{?}{12}$ $\frac{1 \times 2}{6 \times 2} = \frac{?}{12}$

> I used the product of the denominators, 24.
> $\frac{1 \times 6}{4 \times 6} = \frac{6}{24}$ $\frac{1 \times 4}{6 \times 4} = \frac{4}{24}$

4-2 Adding Unlike Fractions

5 Add $\frac{3}{10}$ and $\frac{5}{6}$. Express the answer in simplest form.

I used the least common multiple.
$\frac{3 \times 3}{10 \times 3} = \frac{?}{30}$ $\frac{5 \times 5}{6 \times 5} = \frac{?}{30}$

I used the product of the denominators.
$\frac{3 \times 6}{10 \times 6} = \frac{?}{60}$ $\frac{5 \times 10}{6 \times 10} = \frac{?}{60}$

$\frac{3}{10} + \frac{5}{6} = \boxed{}\frac{\boxed{}}{\boxed{}}$

6 Add. Express each answer in simplest form.

(a) $\frac{1}{2} + \frac{1}{7}$

(b) $\frac{1}{6} + \frac{3}{10}$

(c) $\frac{1}{10} + \frac{2}{5}$

(d) $\frac{2}{3} + \frac{5}{8}$

(e) $\frac{9}{6} + \frac{6}{9}$

(f) $\frac{5}{3} + \frac{3}{2}$

(g) $\frac{1}{2} + \frac{1}{3} + \frac{1}{6}$

(h) $\frac{3}{10} + \frac{4}{5} + \frac{3}{4}$

7 Mei ran $\frac{3}{4}$ mi on Monday, $\frac{1}{2}$ mi on Tuesday, and $\frac{1}{3}$ mi on Wednesday. How far did she run on the three days altogether?

Exercise 2 • page 76

Lesson 3
Subtracting Unlike Fractions

Think

Emma had $\frac{3}{4}$ lb of butter. She used $\frac{1}{3}$ lb to make butter cookies. How many pounds of butter does she have left?

Will the answer be greater than or less than $\frac{1}{2}$ lb?

Learn

$\frac{3}{4} - \frac{1}{3}$

12 is a common multiple of 3 and 4.

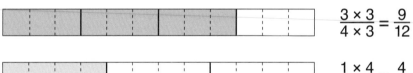

$\frac{3 \times 3}{4 \times 3} = \frac{9}{12}$

$\frac{1 \times 4}{3 \times 4} = \frac{4}{12}$

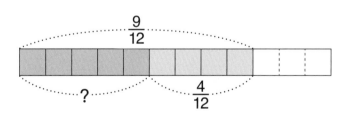

$\frac{3}{4} - \frac{1}{3} = \frac{9}{12} - \frac{4}{12} = \frac{5}{12}$

She has _____ lb of butter left.

Do

1 Subtract $\frac{1}{4}$ from $\frac{5}{8}$.

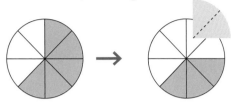

$\frac{1 \times 2}{4 \times 2} = \frac{?}{8}$

$\frac{5}{8} - \frac{1}{4} = \frac{5}{8} - \frac{\square}{8} = \frac{\square}{\square}$

2 Subtract $\frac{1}{2}$ from $\frac{3}{5}$.

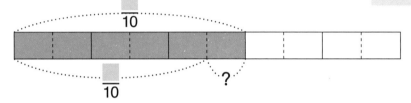

$\frac{3 \times 2}{5 \times 2} = \frac{?}{10} \qquad \frac{1 \times 5}{2 \times 5} = \frac{?}{10}$

$\frac{3}{5} - \frac{1}{2} = \frac{\square}{10} - \frac{\square}{10} = \frac{\square}{\square}$

3 Subtract $\frac{2}{3}$ from $\frac{7}{5}$.

$\frac{7 \times 3}{5 \times 3} = \frac{?}{15} \qquad \frac{2 \times 5}{3 \times 5} = \frac{?}{15}$

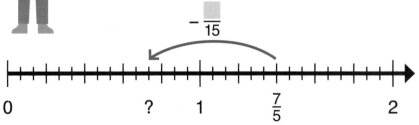

$\frac{7}{5} - \frac{2}{3} = \frac{\square}{15} - \frac{\square}{15} = \frac{\square}{\square}$

4 Subtract $\frac{5}{6}$ from $\frac{8}{9}$. Express the answer in simplest form.

$$\frac{8}{9} - \frac{5}{6} = \frac{\square}{\square}$$

Which common denominator will I use?

$$\frac{8 \times 6}{9 \times 6} = \frac{48}{54} \qquad \frac{5 \times 9}{6 \times 9} = \frac{45}{54}$$

$$\frac{8 \times 2}{9 \times 2} = \frac{16}{18} \qquad \frac{5 \times 3}{6 \times 3} = \frac{15}{18}$$

5 Subtract. Express each answer in simplest form.

(a) $\frac{4}{5} - \frac{2}{3}$

(b) $\frac{5}{6} - \frac{3}{10}$

(c) $\frac{4}{7} - \frac{1}{2}$

(d) $\frac{5}{2} - \frac{5}{8}$

(e) $\frac{5}{4} - \frac{7}{8}$

(f) $1 - \frac{9}{15}$

(g) $\frac{7}{8} - \frac{1}{4} - \frac{1}{2}$

(h) $1 - \frac{1}{2} - \frac{1}{3} - \frac{1}{6}$

6 Franco had 2 lb of flour. He used $\frac{3}{4}$ lb to make bread and $\frac{2}{5}$ lb to make muffins. How many pounds of flour does he have left?

Exercise 3 • page 79

Lesson 4
Practice A

1 Divide. Express each answer in simplest form.

(a) 25 ÷ 7

(b) 32 ÷ 6

(c) 45 ÷ 10

(d) 93 ÷ 13

(e) 460 ÷ 6

(f) 1,000 ÷ 80

2 Express each fraction as a whole number or mixed number in simplest form.

(a) $\frac{20}{4}$

(b) $\frac{21}{14}$

(c) $\frac{40}{6}$

(d) $\frac{18}{4}$

(e) $\frac{97}{8}$

(f) $\frac{570}{12}$

3 Add or subtract. Express each answer in simplest form.

(a) $\frac{3}{8} + \frac{1}{6}$

(b) $\frac{5}{9} + \frac{3}{4}$

(c) $\frac{4}{5} + \frac{7}{10}$

(d) $\frac{5}{6} - \frac{5}{10}$

(e) $\frac{7}{8} - \frac{5}{6}$

(f) $\frac{3}{4} - \frac{3}{5}$

(g) $\frac{7}{10} + \frac{3}{4}$

(h) $\frac{7}{8} - \frac{2}{5}$

4 Find the values. Express each answer in simplest form.

(a) $\frac{5}{6} - \frac{2}{3} - \frac{1}{9}$

(b) $\frac{5}{6} - (\frac{2}{3} - \frac{1}{9})$

(c) $\frac{9}{10} - \frac{1}{5} + \frac{1}{2}$

(d) $\frac{9}{10} - (\frac{1}{5} + \frac{1}{2})$

5 A 25 m long rope is cut into 4 equal pieces. How long is each piece in meters?

6 A baker has 38 kg of flour. He wants to put all of the flour equally into 5 tins. How many kilograms of flour should he put in each tin?

7 A building is 112 ft tall. It is 12 times as tall as a fire truck. How tall is the fire truck in feet?

8 Charlotte spent $\frac{2}{5}$ of her money on a book, $\frac{1}{4}$ of her money on a notebook, and $\frac{1}{10}$ of her money on a pen. What fraction of her money did she spend altogether?

9 Theodore spent $\frac{1}{3}$ of his money on a coat, $\frac{1}{4}$ of his money on shoes, and $\frac{1}{6}$ of his money on a shirt. What fraction of his money does he have left?

Exercise 4 • page 82

Lesson 5
Adding Mixed Numbers — Part 1

Think

Sofia is making granola. She needs $1\frac{1}{2}$ cups of chopped almonds and $\frac{2}{3}$ cups of chopped walnuts. How many cups of chopped nuts does she need to make the granola?

Will the answer be more or less than 2 cups?

Learn

$1\frac{1}{2} + \frac{2}{3}$

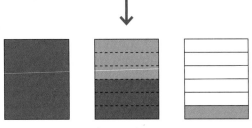

We need to have equal-sized units. 6 is a common multiple of both 2 and 3.

Method 1

$1\frac{1}{2} + \frac{2}{3} = 1\frac{3}{6} + \frac{4}{6}$

$= 1\frac{7}{6}$

$= 2\frac{1}{6}$

$1\frac{7}{6} = 1 + \frac{6}{6} + \frac{1}{6} = 2 + \frac{1}{6} = 2\frac{1}{6}$

Method 2

$1\frac{1}{2} + \frac{2}{3} = 1\frac{3}{6} + \frac{4}{6}$

$= 1\frac{3}{6} + \frac{3}{6} + \frac{1}{6}$

$= 2 + \frac{1}{6}$

$= 2\frac{1}{6}$

I made the next whole number first.

$1\frac{3}{6} + \frac{4}{6}$

$\frac{3}{6} \quad \frac{1}{6}$

Method 3

$1\frac{1}{2} + \frac{2}{3} = \frac{3}{2} + \frac{2}{3}$

$= \frac{9}{6} + \frac{4}{6}$

$= \frac{13}{6}$

$= 2\frac{1}{6}$

I expressed the mixed number as an improper fraction first.

She needs _____ cups of chopped nuts.

Do

1 Add $1\frac{1}{3}$ and $\frac{1}{4}$.

$1\frac{1}{3} + \frac{1}{4} = \frac{4}{3} + \frac{1}{4}$

$= \frac{\square}{12} + \frac{\square}{12}$

$= \frac{\square}{12}$

$= 1\frac{\square}{12}$

$1\frac{1}{3} = \frac{4}{3}$

2 Add $3\frac{1}{3}$ and $\frac{4}{5}$.

$3\frac{1}{3} + \frac{4}{5} = 3\frac{\square}{15} + \frac{\square}{15}$

$= 3\frac{\square}{15}$

$= 4\frac{\square}{15}$

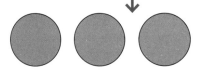

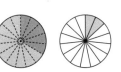

$3\frac{17}{15} = 3 + \frac{15}{15} + \frac{2}{15} = 4 + \frac{2}{15}$

3 Add $\frac{4}{5}$ and $1\frac{7}{10}$.

$\frac{4}{5} + 1\frac{7}{10} = \frac{\square}{10} + 1\frac{7}{10}$

$= 1\frac{\square}{10}$

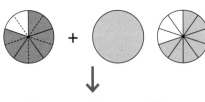

$= 2\frac{\square}{10}$

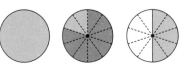

$= 2\frac{\square}{2}$

4-5 Adding Mixed Numbers — Part 1

4 Add $1\frac{3}{4}$ and $\frac{5}{8}$.

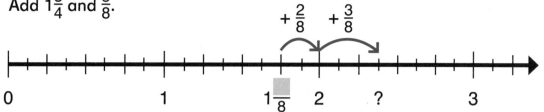

$1\frac{3}{4} + \frac{5}{8} = 1\frac{6}{8} + \frac{5}{8}$

$\phantom{1\frac{3}{4} + \frac{5}{8}} = \square + \frac{3}{8}$

$\phantom{1\frac{3}{4} + \frac{5}{8}} = \square\frac{\square}{\square}$

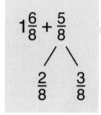

5 Add. Express the answer in simplest form.

(a) $4\frac{1}{8} + \frac{3}{8}$

(b) $1\frac{1}{4} + \frac{1}{2}$

(c) $8\frac{3}{4} + \frac{3}{10}$

(d) $6\frac{4}{5} + \frac{1}{4}$

(e) $\frac{3}{9} + 4\frac{1}{2}$

(f) $\frac{3}{4} + 7\frac{5}{6}$

(g) $\frac{1}{7} + \frac{1}{3} + 9\frac{2}{3}$

(h) $2\frac{5}{9} + \frac{1}{3} + \frac{1}{2}$

6 Hazel used $2\frac{1}{2}$ cups of water, $\frac{1}{3}$ cup of orange juice, and $\frac{1}{4}$ cup of lemon juice to make a refreshing drink. How many cups of liquid did she use altogether?

Exercise 5 • page 85

Lesson 6
Adding Mixed Numbers — Part 2

Think

To make a cherry tart, Emma uses $1\frac{2}{3}$ cups of sweet cherries and $1\frac{1}{2}$ cups of sour cherries. How many cups of cherries does she use?

It will be a little more than 3 cups.

Learn

$1\frac{1}{2} + 1\frac{2}{3}$

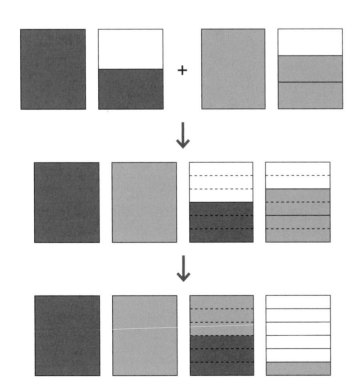

A common multiple of 2 and 3 is...

Method 1

$$1\frac{1}{2} + 1\frac{2}{3} = 2\frac{1}{2} + \frac{2}{3}$$

$$= 2\frac{3}{6} + \frac{4}{6}$$

$$= 2\frac{7}{6}$$

$$= 3\frac{1}{6}$$

$1\frac{3}{6} \xrightarrow{+1} 2\frac{3}{6} \xrightarrow{+\frac{4}{6}} ?$

Method 2

$$1\frac{1}{2} + 1\frac{2}{3} = 1 + \frac{1}{2} + 1 + \frac{2}{3}$$

$$= (1 + 1) + \left(\frac{3}{6} + \frac{4}{6}\right)$$

$$= 2 + \frac{7}{6}$$

$$= 2 + 1 + \frac{1}{6}$$

$$= 3\frac{1}{6}$$

I added the whole number and fraction parts separately.

Method 3

$$1\frac{1}{2} + 1\frac{2}{3} = \frac{3}{2} + \frac{5}{3}$$

$$= \frac{9}{6} + \frac{10}{6}$$

$$= \frac{19}{6}$$

$$= 3\frac{1}{6}$$

I expressed both fractions as improper fractions first.

She uses _____ cups of cherries.

Do

① Add $2\frac{1}{4}$ and $1\frac{2}{3}$.

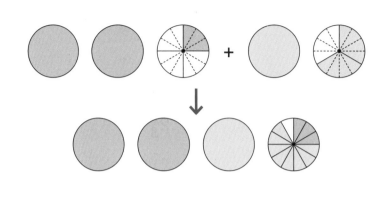

$2\frac{1}{4} + 1\frac{2}{3} = 3\frac{1}{4} + \frac{2}{3}$

$= 3\frac{\square}{12} + \frac{\square}{12}$

$= 3\frac{\square}{12}$

② Add $1\frac{1}{4}$ and $1\frac{5}{6}$.

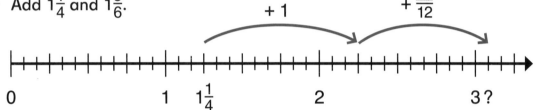

$1\frac{1}{4} + 1\frac{5}{6} = 2\frac{1}{4} + \frac{5}{6}$

$= 2\frac{\square}{12} + \frac{\square}{12}$

$= 3\frac{\square}{12}$

③ Add $5\frac{5}{6}$ and $2\frac{2}{3}$.

$5\frac{5}{6} + 2\frac{2}{3} = 5\frac{5}{6} + 2\frac{\square}{6}$

$= 7\frac{\square}{6}$

$= 8\frac{\square}{6}$

$= 8\frac{\square}{2}$

4. Add. Express each answer in simplest form.

(a) $5\frac{7}{12} + 2$

(b) $3\frac{1}{8} + 1\frac{3}{8}$

(c) $3\frac{1}{3} + 3\frac{6}{9}$

(d) $6\frac{1}{4} + 4\frac{3}{10}$

(e) $1\frac{4}{5} + 2\frac{1}{3}$

(f) $5\frac{1}{2} + 3\frac{5}{6}$

(g) $6\frac{5}{6} + 5\frac{7}{10}$

(h) $7\frac{3}{10} + 2\frac{2}{3}$

(i) $1\frac{1}{2} + 3\frac{1}{4} + 2\frac{1}{2}$

(j) $5\frac{5}{6} + 3\frac{4}{9} + 2\frac{2}{3}$

5.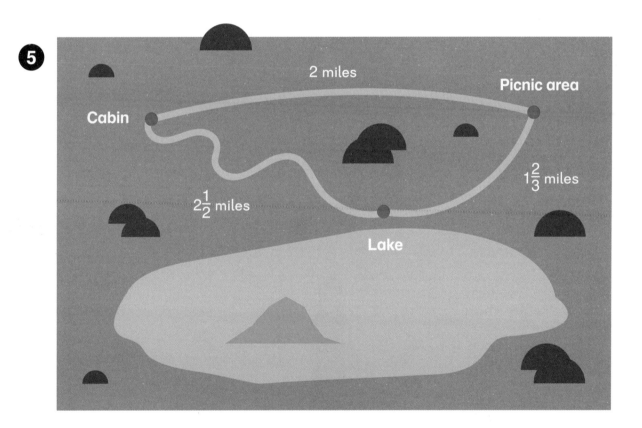

Emma and her dad walked from the cabin to the lake. After that, they walked to the picnic area, then took the shortest route back to the cabin. How far did they walk?

Exercise 6 • page 87

Lesson 7
Subtracting Mixed Numbers — Part 1

Think

Alex had $2\frac{2}{5}$ L of milk. He used $\frac{1}{2}$ L to make milk shakes. How much milk does he have left?

Learn

$2\frac{2}{5} - \frac{1}{2}$

Method 1

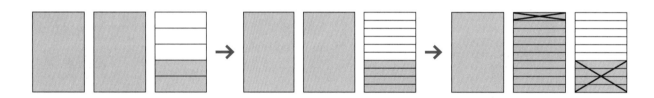

$\frac{5}{10}$ is greater than $\frac{4}{10}$.

$2\frac{4}{10} = 1 + \frac{10}{10} + \frac{4}{10} = 1 + \frac{?}{10}$

$2\frac{2}{5} - \frac{1}{2} = 2\frac{4}{10} - \frac{5}{10}$

$= 1\frac{14}{10} - \frac{5}{10}$

$= 1\frac{9}{10}$

Method 2

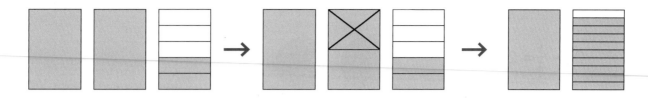

$2\frac{2}{5} - \frac{1}{2} = 2 - \frac{1}{2} + \frac{2}{5}$

$= 1\frac{1}{2} + \frac{2}{5}$

$= 1\frac{5}{10} + \frac{4}{10}$

$= 1\frac{9}{10}$

I subtracted $\frac{1}{2}$ from 2 first.

Method 3

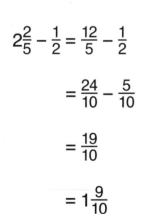

I changed the mixed number to an improper fraction.

$2\frac{2}{5} - \frac{1}{2} = \frac{12}{5} - \frac{1}{2}$

$= \frac{24}{10} - \frac{5}{10}$

$= \frac{19}{10}$

$= 1\frac{9}{10}$

He has _____ L of milk left.

Do

1 Subtract $\frac{1}{4}$ from $2\frac{1}{3}$.

$2\frac{1}{3} - \frac{1}{4} = 2\frac{\square}{12} - \frac{\square}{12}$

$= \square\frac{\square}{12}$

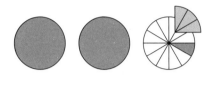

2 Subtract $\frac{3}{5}$ from $2\frac{1}{2}$.

$2\frac{1}{2} - \frac{3}{5} = 2\frac{5}{10} - \frac{6}{10}$

$= 1\frac{\square}{10} - \frac{6}{10}$

$= \square\frac{\square}{\square}$

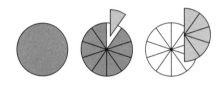

$2\frac{5}{10} = 1 + \frac{10}{10} + \frac{5}{10} = 1\frac{?}{10}$

3 Subtract $\frac{5}{6}$ from $2\frac{1}{3}$.

$2\frac{1}{3} - \frac{5}{6} = 2 - \frac{5}{6} + \frac{1}{3}$

$= 1\frac{1}{6} + \frac{1}{3}$

$= 1\frac{1}{6} + \frac{2}{6}$

$= 1\frac{\square}{6}$

$= \square\frac{\square}{\square}$

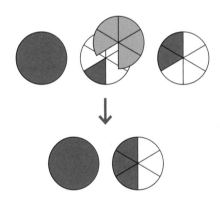

I subtracted $\frac{5}{6}$ from 2 first.

4. Subtract $\frac{2}{3}$ from $1\frac{1}{2}$.

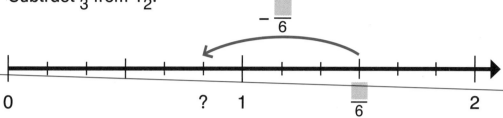

$1\frac{1}{2} - \frac{2}{3} = \frac{3}{2} - \frac{2}{3}$

$= \frac{\square}{6} - \frac{\square}{6}$

$= \frac{\square}{6}$

5. Subtract. Express each answer in simplest form.

(a) $5 - \frac{4}{5}$

(b) $3\frac{5}{7} - \frac{3}{7}$

(c) $2\frac{1}{4} - \frac{3}{8}$

(d) $5\frac{1}{2} - \frac{1}{3}$

(e) $8\frac{1}{5} - \frac{3}{4}$

(f) $7\frac{3}{8} - \frac{7}{10}$

(g) $6\frac{1}{6} - \frac{3}{10}$

(h) $3 - \frac{1}{2} - \frac{4}{5}$

6. Dion wants to run 5 miles this week. He ran $\frac{4}{5}$ miles on Monday and $\frac{9}{10}$ miles on Tuesday. How many more miles does he need to run?

Exercise 7 • page 90

Lesson 8
Subtracting Mixed Numbers — Part 2

Think

Sofia's mom had $3\frac{1}{2}$ cups of coffee beans. She used $1\frac{2}{3}$ cups to make coffee. How many cups of coffee beans does she have left?

Learn

$3\frac{1}{2} - 1\frac{2}{3}$

Method 1

$3\frac{1}{2} \xrightarrow{-1} 2\frac{1}{2} \xrightarrow{-\frac{2}{3}} ?$

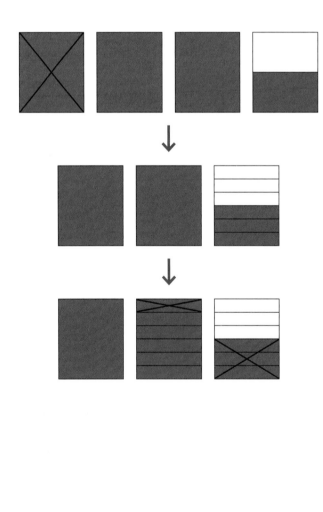

$3\frac{1}{2} - 1\frac{2}{3} = 2\frac{1}{2} - \frac{2}{3}$

$= 2\frac{3}{6} - \frac{4}{6}$

$= 1\frac{9}{6} - \frac{4}{6}$

$= 1\frac{5}{6}$

To calculate the step with $2\frac{3}{6} - \frac{4}{6}$, we could also do $2 - \frac{4}{6} + \frac{3}{6} = 1\frac{2}{6} + \frac{3}{6}$.

Method 2

I changed the mixed numbers to improper fractions first.

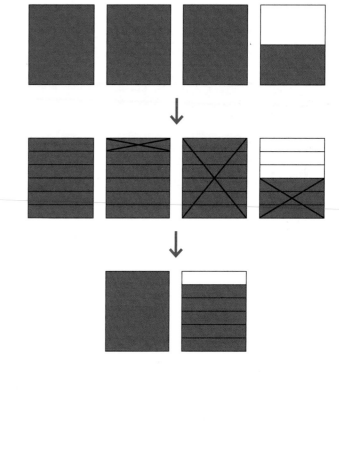

$3\frac{1}{2} - 1\frac{2}{3} = 3\frac{3}{6} - 1\frac{4}{6}$

$= \frac{21}{6} - \frac{10}{6}$

$= \frac{11}{6}$

$= 1\frac{5}{6}$

She has _____ cups of coffee beans left.

Do

1 Subtract $2\frac{1}{5}$ from $4\frac{1}{2}$.

$4\frac{1}{2} - 2\frac{1}{5} = 2\frac{1}{2} - \frac{1}{5}$

$= 2\frac{\boxed{}}{10} - \frac{\boxed{}}{10}$

$= \boxed{}\frac{\boxed{}}{\boxed{}}$

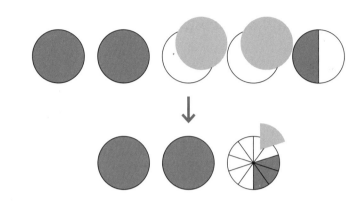

2 Subtract $1\frac{3}{4}$ from $2\frac{1}{3}$.

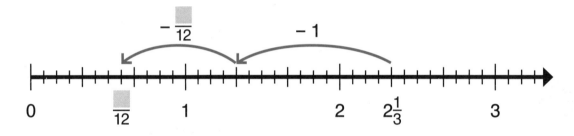

$2\frac{1}{3} - 1\frac{3}{4} = 1\frac{1}{3} - \frac{3}{4}$

$= 1\frac{\boxed{}}{12} - \frac{9}{12}$

$= \frac{\boxed{}}{12} - \frac{9}{12}$

$= \frac{\boxed{}}{12}$

I subtracted the whole number first.

3 Subtract $1\frac{2}{3}$ from $2\frac{1}{2}$.

$2\frac{1}{2} - 1\frac{2}{3} = \frac{\boxed{}}{2} - \frac{\boxed{}}{3} = \frac{\boxed{}}{6} - \frac{\boxed{}}{6} = \frac{\boxed{}}{6}$

I changed the mixed numbers to improper fractions first.

4 $9\frac{1}{10} - 4\frac{5}{6} = 5\frac{1}{10} - \frac{5}{6}$

$= 5\frac{3}{30} - \frac{\square}{30}$

$= 4\frac{\square}{30} - \frac{\square}{30}$

$= 4\frac{\square}{30}$

$= \square\frac{\square}{\square}$

$5\frac{3}{30} = 4 + 1\frac{3}{30}$
$= 4 + \frac{30}{30} + \frac{3}{30}$
$= 4\frac{33}{30}$

5 Subtract. Express each answer in simplest form.

(a) $5\frac{7}{8} - 2\frac{1}{4}$

(b) $7\frac{1}{2} - 3\frac{3}{10}$

(c) $2\frac{5}{6} - 1\frac{1}{4}$

(d) $2\frac{1}{3} - 1\frac{1}{5}$

(e) $8\frac{1}{4} - \frac{7}{10}$

(f) $9\frac{1}{6} - 2\frac{5}{9}$

(g) $7 - \frac{1}{3} - 3\frac{4}{5}$

(h) $8\frac{3}{8} - 1\frac{3}{10} - \frac{3}{4}$

6 Mei had $5\frac{1}{2}$ m of ribbon. She used $1\frac{4}{5}$ m to tie cake boxes and $\frac{3}{4}$ m for a balloon. How many meters of ribbon does she have left?

Exercise 8 • page 92

Lesson 9
Practice B

1 Add or subtract. Express each answer in simplest form.

(a) $1\frac{3}{4} + \frac{4}{5}$

(b) $\frac{1}{5} + 2\frac{3}{10}$

(c) $4\frac{5}{9} + 2\frac{1}{3}$

(d) $1\frac{3}{8} - \frac{1}{2}$

(e) $3\frac{2}{3} - 1\frac{3}{7}$

(f) $6\frac{3}{10} - 4\frac{5}{6}$

(g) $6\frac{5}{6} + 3\frac{3}{8}$

(h) $9\frac{3}{4} - 3\frac{8}{9}$

2 Find the values. Express each answer in simplest form.

(a) $2\frac{1}{2} + \frac{3}{4} - \frac{5}{8}$

(b) $1\frac{1}{2} - \frac{2}{3} + 5\frac{5}{6}$

(c) $2\frac{2}{3} - (4\frac{1}{2} - 1\frac{5}{6})$

(d) $2\frac{1}{4} - (\frac{3}{4} + \frac{3}{10})$

(e) $12 - (4\frac{1}{2} - 1\frac{3}{8}) - 3\frac{1}{4}$

(f) $(6\frac{1}{3} - 1\frac{3}{4}) + (6\frac{1}{3} - 1\frac{3}{4})$

3 Shanice had a 2 L bottle of water. She drank $1\frac{2}{3}$ L of water. How much water does she have left?

4 Rope A is $8\frac{3}{5}$ m long. Rope B is $5\frac{3}{10}$ m long. How much longer is Rope A than Rope B?

5 Sara is walking from her house to the library. The library is $1\frac{1}{2}$ miles from her house. She has already walked $\frac{3}{8}$ miles. How much farther does she have to go?

6 Parker ran $2\frac{3}{5}$ miles on Friday, $3\frac{1}{2}$ miles on Saturday, and $2\frac{1}{10}$ miles on Sunday. How far did he run over the three days?

7 Madison spent $2\frac{2}{3}$ hours working on a project on Friday and $2\frac{1}{2}$ hours working on Saturday. She finished the project on Sunday. If the project took $6\frac{3}{4}$ hours, how many hours did she spend working on Sunday?

8 Wyatt has 16 L of orange juice and 12 L of peach juice. He poured all of the juice into 5 pitchers. Each pitcher has an equal amount of each kind of juice. How many liters of juice are in each pitcher?

9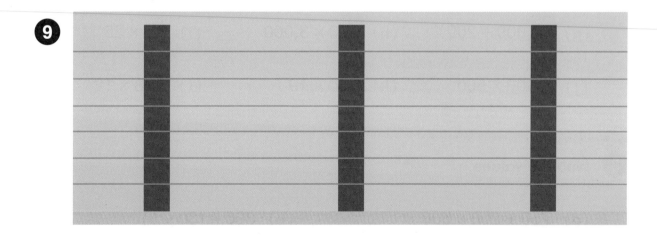

Dennis built a wire fence 168 feet long. He used 17 fence posts and spaced them equally. There is one fence post at each end of the fence. How many feet apart was each fence post?

Exercise 9 • page 94

Review 1

1 | 673,895,245 | 98,576,056 | 667,300,006 | 97,203,450 | 700,003,000 |

(a) Put the numbers in order from least to greatest.

(b) What is the value of the digit 7 in each of the numbers?

(c) Write each of the numbers in words.

2 Find the values. Use mental calculation.

(a) 2,055 × 100   (b) 9,850 ÷ 10   (c) 850,000 ÷ 1,000

(d) 4,000 × 7,000   (e) 6,400 ÷ 80   (f) 15,000 × 40

(g) 32,000 ÷ 200   (h) 270 × 3,000   (i) 99 × 25

(j) 6,400 × 500   (k) 320 × 49   (l) 998 × 12

3 Find the values.

(a) 750 + 200 × 500   (b) 300 × (30 − 5)

(c) 900 − 15 ÷ 3 × 2   (d) 640 ÷ (264 − 256)

(e) 85 × 30 − 56 ÷ 7 + 30   (f) 9 × (12 − 45 ÷ 5) + 18 − 2 × 5

4 Divide. Express each answer as a mixed number in simplest form.

(a) 5 ÷ 4   (b) 24 ÷ 7   (c) 52 ÷ 12   (d) 140 ÷ 40

---

5 Express each fraction as a whole number or mixed number in simplest form.

(a) $\frac{25}{6}$ (b) $\frac{48}{18}$ (c) $\frac{300}{200}$ (d) $\frac{360}{90}$

6 Estimate and then find the actual values, expressed in simplest form.

(a) 98 × 654   (b) 86 × 243   (c) 6,263 × 47

(d) 81 × 21 × 15   (e) 86 ÷ 23   (f) 287 ÷ 59

(g) 4,864 ÷ 36   (h) 34,845 ÷ 15   (i) 24,321 ÷ 75

7 Find the values. Express each answer in simplest form.

(a) $\frac{7}{8} - \frac{3}{5}$ (b) $\frac{5}{9} + \frac{5}{6}$

(c) $\frac{9}{4} - \frac{3}{2} - \frac{1}{3}$ (d) $\frac{1}{2} + \frac{3}{7} + \frac{2}{3}$

(e) $\frac{5}{6} + 4\frac{1}{3}$ (f) $5\frac{2}{3} - \frac{1}{5}$

(g) $5\frac{1}{5} - 4\frac{1}{7}$ (h) $6\frac{3}{5} - 3\frac{7}{8}$

(i) $5\frac{3}{4} - (1\frac{1}{2} + 2\frac{1}{6})$ (j) $(4\frac{1}{3} - 1\frac{1}{2}) + (\frac{9}{2} - 3\frac{1}{4})$

Review 1

8 Liam and Pekelo together have $8,500. Liam has $750 more than Pekelo. How much money does each person have?

9 Grapefruits are sold at 4 for $5. After buying 20 grapefruits, Selena had $18 left. How much money did she have at first?

10 2 watermelons and 3 cantaloupes cost $35. 4 watermelons and 3 cantaloupes cost $55. How much does one cantaloupe cost?

11 Dana had an equal number of red beads and yellow beads. After using 500 red beads to make necklaces, she had 3 times as many yellow beads as red beads. How many yellow beads does she have?

12 32 lb of quinoa is put equally into 12 bags. How many pounds of quinoa are in each bag?

13 A bag has $3\frac{1}{2}$ lb of flour. After using some of it to make bread there was $\frac{3}{8}$ lb of flour left. How many pounds of flour were used?

14 On Monday, Jason ran $3\frac{1}{2}$ miles in the morning and $1\frac{3}{4}$ miles in the afternoon. On Tuesday, he ran $\frac{7}{8}$ miles in the morning and $4\frac{1}{3}$ miles in the afternoon. On which day did he run farther? How much farther?

Exercise 10 • page 97

Chapter 5

Multiplication of Fractions

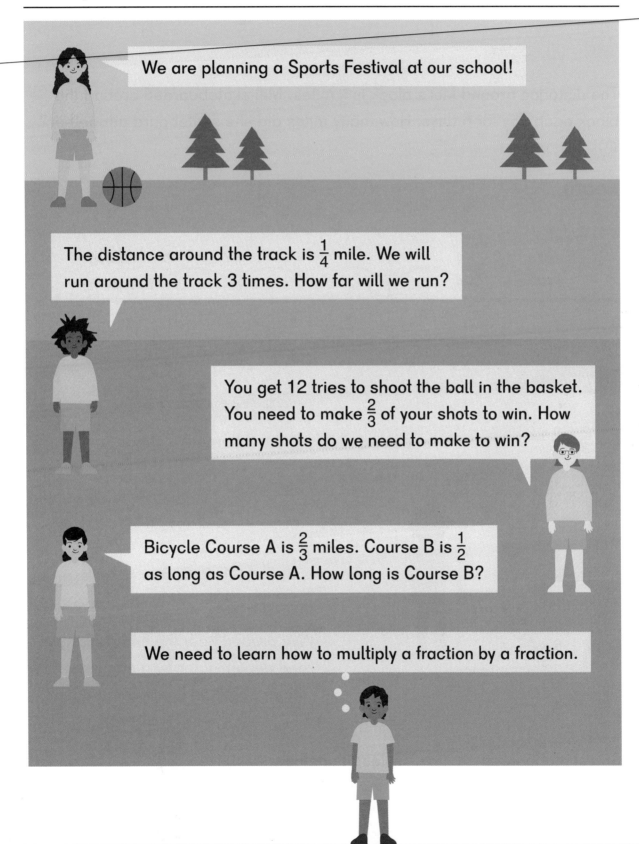

We are planning a Sports Festival at our school!

The distance around the track is $\frac{1}{4}$ mile. We will run around the track 3 times. How far will we run?

You get 12 tries to shoot the ball in the basket. You need to make $\frac{2}{3}$ of your shots to win. How many shots do we need to make to win?

Bicycle Course A is $\frac{2}{3}$ miles. Course B is $\frac{1}{2}$ as long as Course A. How long is Course B?

We need to learn how to multiply a fraction by a fraction.

Lesson 1
Multiplying a Fraction by a Whole Number

Think

The distance around Mei's block is $\frac{2}{3}$ miles. Mei skateboarded around the block each day for 6 days. How many miles did she skateboard altogether?

Learn

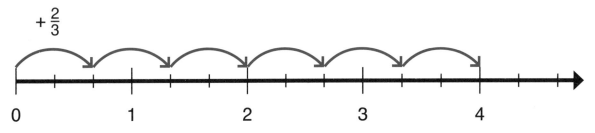

Method 1

$6 \times \frac{2}{3} = \frac{6 \times 2}{3}$

$= \frac{12}{3}$

$= 4$

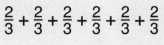

$\frac{2}{3} + \frac{2}{3} + \frac{2}{3} + \frac{2}{3} + \frac{2}{3} + \frac{2}{3}$

6 × 2 thirds = ? thirds

Factors of 3 ⟶ 1, **3**
Factors of 6 ⟶ 1, 2, **3**, 6

Method 2

$6 \times \frac{2}{3} = \frac{\overset{2}{\cancel{6}} \times 2}{\underset{1}{\cancel{3}}}$

$= 4$

We divided both the numerator and denominator by a common factor to get an equivalent fraction.

She skateboarded _____ miles altogether.

Do

① Find the product of 2 and $\frac{3}{7}$.

2 × 3 sevenths = ? sevenths

$$2 \times \frac{3}{7} = \frac{2 \times 3}{7}$$

$$= \frac{\square}{7}$$

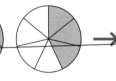

② Find the product of 5 and $\frac{3}{4}$.

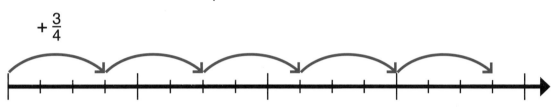

$$5 \times \frac{3}{4} = \frac{5 \times 3}{4}$$

$$= \frac{\square}{4}$$

$$= \square \frac{\square}{4}$$

5 × 3 fourths = ? fourths

③ Find the product of 6 and $\frac{4}{9}$.

$$6 \times \frac{4}{9} = \frac{\overset{2}{\cancel{6}} \times 4}{\underset{3}{\cancel{9}}} = \frac{8}{3}$$

$$= \square \frac{\square}{\square}$$

6 and 9 have a common factor of 3, so we can simplify the calculation.

5-1 Multiplying a Fraction by a Whole Number

4 Find the product of 8 and $\frac{5}{12}$.

$8 \times \frac{5}{12} = \frac{\overset{2}{\cancel{8}} \times 5}{\underset{3}{\cancel{12}}}$

Factors of 8 → **1, 2, 4, 8**
Factors of 12 → **1, 2, 3, 4, 6, 12**

$= \frac{\square}{3}$

$= 3\frac{\square}{3}$

We can also simplify in more than one step.

$8 \times \frac{5}{12}$

5 Find the values. Express each answer in simplest form.

(a) $4 \times \frac{2}{9}$ (b) $5 \times \frac{3}{7}$ (c) $6 \times \frac{3}{8}$

(d) $9 \times \frac{5}{9}$ (e) $5 \times \frac{2}{3}$ (f) $20 \times \frac{3}{8}$

(g) $6 \times \frac{7}{10}$ (h) $6 \times \frac{5}{24}$ (i) $50 \times \frac{3}{100}$

6 The distance around a track is $\frac{2}{5}$ km. A race was 3 laps around the track. How long was the race in kilometers?

7 $\frac{9}{10}$ gallons of water flows from a running faucet every minute. How many gallons of water will flow from the faucet in 30 minutes?

8 Write a word problem for $7 \times \frac{2}{3}$.

Exercise 1 • page 103

Lesson 2
Multiplying a Whole Number by a Fraction

Think

Ella took 20 basketball shots. $\frac{3}{5}$ of them went in the basket. How many baskets did she make?

Did I make more or fewer than 20 baskets?

Learn

Method 1

$\frac{1}{5}$ of 20

$\frac{3}{5}$ of 20

$\frac{1}{5}$ of 20 → $\frac{20}{5}$ = 20 ÷ 5 = 4

$\frac{3}{5}$ of 20 → 3 × $\frac{20}{5}$ = 3 × 4 = 12

$\frac{3}{5}$ of 20 = 3 × ($\frac{1}{5}$ of 20)

Method 2

$\frac{3}{5}$ of 20 is the number that is $\frac{3}{5}$ times as much as 20.

5 units ⟶ 20

1 unit ⟶ $\frac{20}{5}$ = 4

3 units ⟶ 3 × $\frac{20}{5}$ = 3 × 4 = 12

Method 3

$\frac{3}{5} \times 20 = \frac{3 \times \cancel{20}^{4}}{\cancel{5}_{1}}$

$= 12$

$\frac{3}{5} \times 20 = 20 \times \frac{3}{5}$
I can use the same method I used when multiplying a whole number by a fraction.

She made _____ baskets.

Do

① Find $\frac{2}{3}$ of 9.

$\frac{1}{3}$ of 9 = 9 ÷ 3

$\frac{2}{3}$ of 9 = 2 × (9 ÷ 3)

$\frac{1}{3} \times 9 = \frac{9}{3}$

$\frac{2}{3} \times 9 = 2 \times \frac{9}{3} = \square$

② (a) Find $\frac{1}{3}$ of 10.

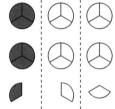

$\frac{1}{3} \times 10 = \frac{\square}{3}$

= $\square \frac{\square}{3}$

$\frac{1}{3}$ of 10 = 10 ÷ 3

(b) Find $\frac{2}{3}$ of 10.

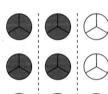

$\frac{2}{3} \times 10 = 2 \times \frac{\square}{3}$

= $\frac{\square}{3}$

= $\square \frac{\square}{\square}$

③ Find $\frac{3}{4}$ of 6.

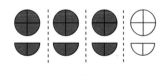

$\frac{3}{4} \times 6 = 3 \times \frac{\square}{4}$

= $\frac{\square}{2}$

= $\square \frac{\square}{\square}$

5-2 Multiplying a Whole Number by a Fraction

4. Find $\frac{3}{4}$ of 12.

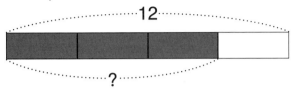

$\frac{3}{4}$ is less than 1 so the product will be less than 12.

(a) 4 units ⟶ 12

1 unit ⟶ $\frac{12}{4}$

3 units ⟶ $3 \times \frac{12}{4} = \square$

$\frac{12}{4} = 3$

(b) $\frac{3}{4} \times 12 = \dfrac{3 \times \cancel{12}^{3}}{\cancel{4}_{1}}$

$= \square$

5. Find the number that is $\frac{4}{3}$ times as large as 12.

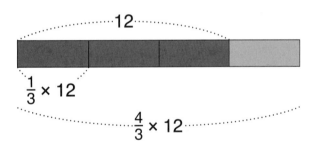

How do we know before calculating that the product will be greater than 12?

(a) 3 units ⟶ 12

1 unit ⟶ $\frac{12}{3} = \square$

4 units ⟶ $4 \times \square = \square$

(b) $\frac{4}{3} \times 12 = \dfrac{4 \times \cancel{12}^{4}}{\cancel{3}_{1}}$

$= \square$

5-2 Multiplying a Whole Number by a Fraction

6 Find $\frac{3}{8}$ of 20.

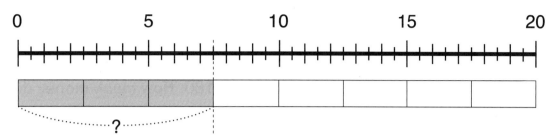

(a) 8 units ⟶ 20

1 unit ⟶ $\frac{20}{8}$

3 units ⟶ $\square \times \frac{\overset{5}{\cancel{20}}}{\underset{2}{\cancel{8}}} = \frac{\square}{2} = \square \frac{\square}{2}$

How do we know the product will be less than 20?

(b) $\frac{3}{\underset{2}{\cancel{8}}} \times \overset{5}{\cancel{20}} = \frac{\square}{2} = \square \frac{\square}{2}$

7 Multiply. Express each answer in simplest form.

(a) $\frac{2}{5} \times 10$ (b) $\frac{5}{2} \times 10$ (c) $\frac{2}{5} \times 2$

(d) $\frac{5}{2} \times 2$ (e) $\frac{5}{6} \times 15$ (f) $\frac{6}{5} \times 15$

(g) $\frac{4}{5} \times 100$ (h) $\frac{5}{4} \times 100$ (i) $\frac{3}{4} \times 60$

(j) $\frac{2}{3} \times 93$ (k) $\frac{5}{6} \times 20$ (l) $\frac{7}{12} \times 8$

8 At a bakery, $\frac{5}{8}$ of a 30-kg bag of flour was used to make bread. How many kilograms of flour were used to make bread?

Exercise 2 • page 105

5-2 Multiplying a Whole Number by a Fraction

Lesson 3
Word Problems — Part 1

Think

Dion has $\frac{3}{4}$ as much money as Alex. Alex has $180. How much money do they have altogether?

Learn

If we show 3 units for Dion and 4 for Alex, then Dion's bar is $\frac{3}{4}$ as long as Alex's bar.

Method 1

4 units ⟶ 180
1 unit ⟶ 180 ÷ 4 = 45
7 units ⟶ 7 × 45 = 315

Method 2

$\frac{3}{4} \times 180 = \frac{3 \times 180}{4} = 135$

135 + 180 = 315

They have $_____ altogether.

Do

1 There are 120 people at the zoo. $\frac{3}{5}$ of them are children and the rest are adults. How many adults are there?

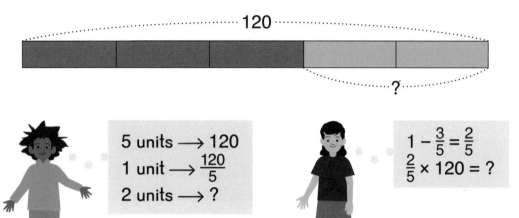

5 units ⟶ 120
1 unit ⟶ $\frac{120}{5}$
2 units ⟶ ?

$1 - \frac{3}{5} = \frac{2}{5}$
$\frac{2}{5} \times 120 = ?$

2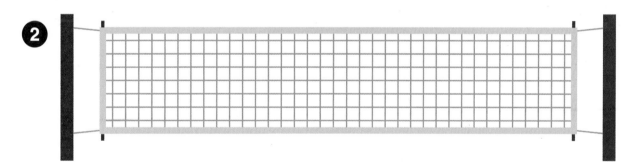

Kiara spent $\frac{3}{8}$ of her money on a volleyball and the rest on a volleyball net. The volleyball net cost $60. How much did the volleyball cost?

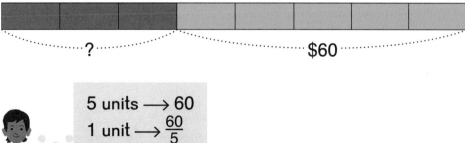

5 units ⟶ 60
1 unit ⟶ $\frac{60}{5}$
3 units ⟶ ?

3

There are 253 dogs at a dog show. There are $\frac{4}{7}$ as many large-breed dogs as small-breed dogs. How many more small-breed dogs are there than large-breed dogs?

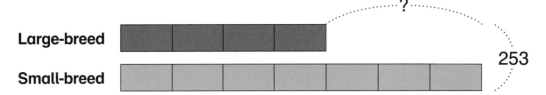

4 Twice as many students are in the chess club as in the fencing club. The number of students in the fencing club is $\frac{1}{5}$ the number of students in the cooking club. There are 42 more students in the cooking club than in the chess club. How many students are in the chess club?

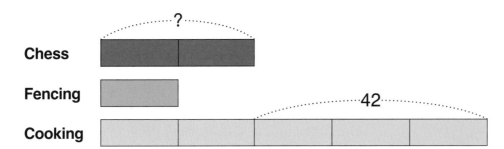

5

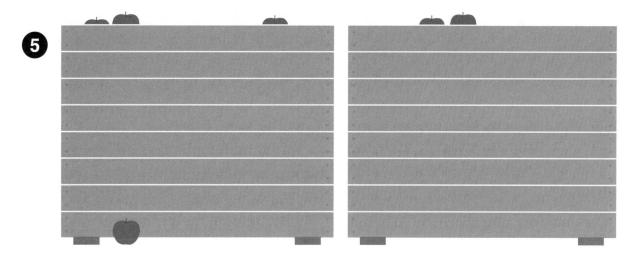

Crate A and Crate B have 156 apples altogether. $\frac{2}{5}$ of the apples in Crate A is equal to $\frac{1}{4}$ of the apples in Crate B. How many apples are in each crate?

6 Aki and Cora had 105 stickers altogether. After Aki used $\frac{1}{3}$ of her stickers and Cora used 25 stickers, they had the same number of stickers left. How many stickers did Cora have at first?

After:

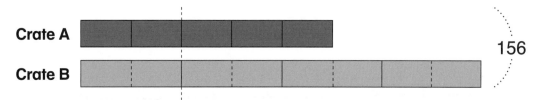

Before:

Exercise 3 • page 107

Lesson 4
Practice A

1 Find the product. Express each answer in simplest form.

(a) $5 \times \frac{2}{3}$

(b) $7 \times \frac{3}{5}$

(c) $4 \times \frac{3}{8}$

(d) $\frac{5}{7} \times 100$

(e) $\frac{7}{9} \times 18$

(f) $\frac{3}{4} \times 70$

(g) $\frac{11}{12} \times 24$

(h) $\frac{5}{14} \times 42$

(i) $\frac{3}{4} \times 75$

(j) $450 \times \frac{3}{5}$

(k) $\frac{1}{3} + \frac{1}{4} \times 12$

(l) $(\frac{1}{2} \times 23) + (\frac{1}{2} \times 17)$

2 There are 40 people at a concert recital. $\frac{3}{5}$ of them are adults and the rest are children. How many children are at the recital?

3 Ximena makes $320 a week working part time. Each week she saves $\frac{1}{4}$ of her money. How much money will she save in 4 weeks?

4 Fadiya spends $\frac{3}{7}$ of her monthly income on rent. Her rent is $1,500 a month. What is her monthly income?

5 Oliver had $250. He spent $\frac{3}{5}$ of it on a suit and $75 on shoes. How much money does he have left?

6. There are 600 beads in a box. $\frac{1}{3}$ of them are red, $\frac{2}{5}$ of them are blue, and the rest are yellow. How many yellow beads are there?

7.

For 3 days, Mari ran $\frac{3}{4}$ miles each day, and for the next 3 days she ran $\frac{2}{3}$ miles each day. How many miles did she run in all?

8. Mila the dog weighs 72 lb. Her puppy Bailey weighs $\frac{1}{5}$ as much as Mila. How many pounds more does Mila weigh than Bailey?

9. There were 352 children at the Sports Festival at first. After $\frac{2}{5}$ of the girls and $\frac{1}{2}$ of the boys left the festival, there were an equal number of boys and girls still at the festival. How many girls were still at the festival?

10. $\frac{2}{3}$ of Maurice's savings is equal to $\frac{4}{5}$ of Ryan's savings. If Maurice saved $10 more than Ryan, how much money did both boys save altogether?

Exercise 4 • page 111

Lesson 5
Multiplying a Fraction by a Unit Fraction

Think

Fold a rectangular paper in fourths horizontally and shade $\frac{1}{4}$ of the paper one color. Then fold the paper in half vertically and shade $\frac{1}{2}$ of the paper another color. What fraction of the paper has been shaded with both colors? What is $\frac{1}{2}$ of $\frac{1}{4}$?

Learn

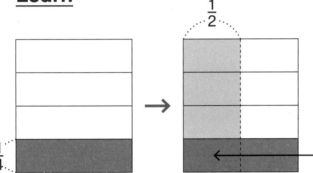

Each fourth is divided into two equal parts. The whole paper is divided into 2 × 4 = 8 equal parts.

$\frac{1}{2}$ of $\frac{1}{4}$ is $\frac{1}{8}$.

$\frac{1}{2}$ of $\frac{1}{4}$ is the number that is $\frac{1}{2}$ times as much as $\frac{1}{4}$. We can write $\frac{1}{2}$ of $\frac{1}{4}$ as $\frac{1}{2} \times \frac{1}{4}$.

$\frac{1}{2} \times \frac{1}{4} = \frac{1 \times 1}{2 \times 4} = \frac{1}{8}$

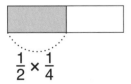

$\frac{1}{2} \times \frac{1}{4}$

Do

1 (a) Find $\frac{1}{3}$ of $\frac{1}{4}$.

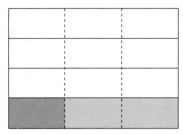

$\frac{1}{3} \times \frac{1}{4} = \frac{1}{\Box}$

(b) Find $\frac{1}{4}$ of $\frac{1}{3}$.

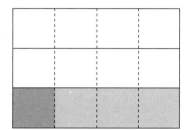

$\frac{1}{4} \times \frac{1}{3} = \frac{1}{\Box}$

The number of parts in the whole rectangle is $3 \times 4 = 12$.

$\frac{1}{4} \times \frac{1}{3} = \frac{1 \times 1}{4 \times 3}$

2 $\frac{3}{4}$ of a garden is planted with roses. $\frac{1}{3}$ of the rose section is planted with white roses. What fraction of the garden is planted with white roses?

$\frac{1}{3} \times \frac{3}{4} = \frac{1 \times 3}{3 \times 4}$

$= \frac{\Box}{12}$

$= \frac{\Box}{4}$

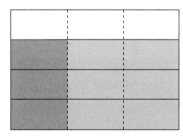

We could also show it this way.

3. (a) Find $\frac{1}{2}$ of $\frac{1}{3}$.

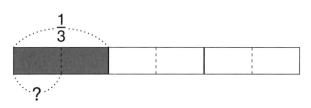

$\frac{1}{2} \times \frac{1}{3} = \frac{1 \times 1}{2 \times 3} = \frac{\square}{\square}$

(b) Find $\frac{1}{2}$ of $\frac{2}{3}$.

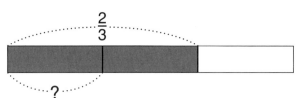

$\frac{1}{2} \times \frac{2}{3} = \frac{1 \times 2}{2 \times 3} = \frac{\square}{6} = \frac{\square}{3}$

$\frac{1}{2}$ of 2 thirds = ? third

4. Alex had $\frac{3}{5}$ L of water. He drank $\frac{1}{6}$ of the water. How much water did he drink in liters?

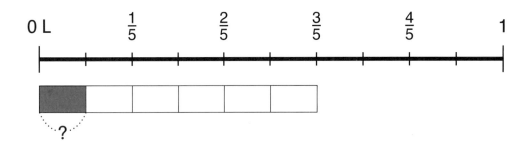

$\frac{1}{6} \times \frac{3}{5} = \frac{1 \times 3}{6 \times 5} = \frac{\square}{30} = \frac{\square}{\square}$

5. Find the values. Express each answer in simplest form.

(a) $\frac{1}{2} \times \frac{1}{5}$ (b) $\frac{1}{2} \times \frac{4}{5}$ (c) $\frac{1}{6} \times \frac{1}{3}$

(d) $\frac{1}{6} \times \frac{2}{3}$ (e) $\frac{1}{3} \times \frac{1}{10}$ (f) $\frac{1}{3} \times \frac{9}{10}$

(g) $\frac{1}{4} \times \frac{4}{7}$ (h) $\frac{1}{5} \times \frac{5}{8}$ (i) $\frac{1}{3} \times \frac{5}{12}$

Exercise 5 • page 115

Lesson 6
Multiplying a Fraction by a Fraction — Part 1

Think

$\frac{3}{5}$ of a field is covered with grass. $\frac{2}{3}$ of the grassy part is used for sports activities. What fraction of the whole field is used for sports activities?

Will the answer be greater than or less than $\frac{3}{5}$?

Learn

Method 1

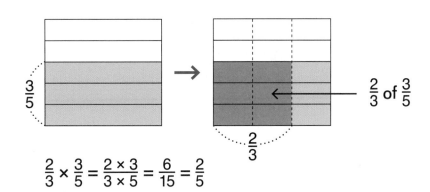

$$\frac{2}{3} \times \frac{3}{5} = \frac{2 \times 3}{3 \times 5} = \frac{6}{15} = \frac{2}{5}$$

Method 2

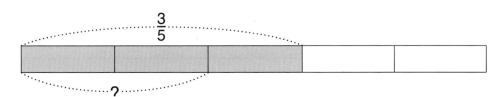

3 units ⟶ $\frac{3}{5}$

1 unit ⟶ $\frac{1}{3} \times \frac{3}{5} = \frac{3}{15} = \frac{1}{5}$

2 units ⟶ $2 \times \frac{1}{5} = \frac{2}{5}$

$\frac{2}{3}$ of $\frac{3}{5}$ = 2 × ($\frac{1}{3}$ of $\frac{3}{5}$)

_____ of the field was used for sports activities.

Do

1 Find $\frac{2}{5}$ of $\frac{2}{3}$.

$\frac{2}{5} \times \frac{2}{3} = \frac{2 \times 2}{5 \times 3}$

$= \frac{\square}{\square}$

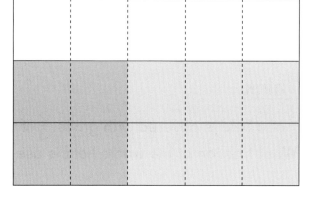

2 Find $\frac{3}{5}$ of $\frac{5}{6}$.

$\frac{3}{5} \times \frac{5}{6} = \frac{3 \times 5}{5 \times 6}$

$= \frac{\square}{30}$

$= \frac{\square}{\square}$

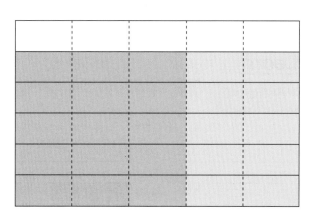

3 Find the value of $\frac{3}{4} \times \frac{4}{5}$.

(a) $\frac{3}{4} \times \frac{4}{5} = \frac{3 \times 4}{4 \times 5}$

$= \frac{\square}{20}$

$= \frac{\square}{5}$

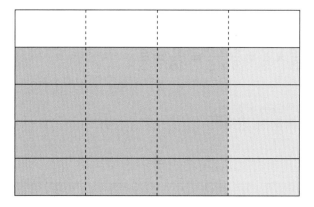

(b) 4 units ⟶ $\frac{4}{5}$

1 unit ⟶ $\frac{1}{4} \times \frac{4}{5} = \frac{4}{20} = \frac{1}{5}$

3 units ⟶ $3 \times \frac{1}{5} = \frac{\square}{5}$

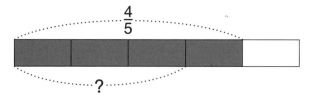

$\frac{3}{4} \times \frac{4}{5} = 3 \times \left(\frac{1}{4} \times \frac{4}{5}\right)$

4 Find the value of $\frac{3}{2} \times \frac{3}{5}$.

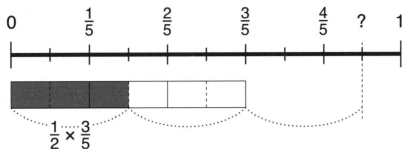

$\frac{3}{2} \times \frac{3}{5} = \frac{3 \times 3}{2 \times 5}$

$= \frac{\square}{\square}$

How do we know the product will be greater than $\frac{3}{5}$?

$\frac{1}{2}$ of $\frac{3}{5} = \frac{3}{10}$

$\frac{3}{2}$ of $\frac{3}{5} = 3 \times \frac{3}{10}$

5 Find the values. Express each answer in simplest form.

(a) $\frac{1}{3} \times \frac{1}{7}$ (b) $\frac{1}{2} \times \frac{5}{6}$ (c) $\frac{4}{5} \times \frac{5}{6}$

(d) $\frac{4}{9} \times \frac{2}{3}$ (e) $\frac{4}{3} \times \frac{4}{5}$ (f) $\frac{3}{2} \times \frac{7}{5}$

6 $\frac{2}{3}$ of a garden is planted with flowers. $\frac{3}{4}$ of the flower section is planted with lilies and the rest is planted with irises.

(a) What fraction of the garden is planted with lilies?

(b) What fraction of the garden is planted with irises?

Exercise 6 • page 117

Lesson 7
Multiplying a Fraction by a Fraction — Part 2

Think

Sofia is swimming across a lake. The lake is $\frac{2}{3}$ km long. She has already swum $\frac{3}{4}$ of the distance. How many kilometers has she swum so far?

Learn

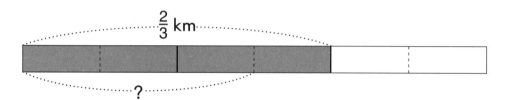

It will be $\frac{3}{4} \times \frac{2}{3}$. How can we simplify this calculation?

2 and 4 have a common factor of 2.
3 and 3 have a common factor of 3.

$$\frac{3}{4} \times \frac{2}{3} = \frac{\cancel{3}^1 \times \cancel{2}^1}{\cancel{4}_2 \times \cancel{3}_1} = \frac{1}{2}$$

$$\frac{3}{4} \times \frac{2}{3} = \frac{3 \times 2}{4 \times 3} = \frac{3 \times 2}{3 \times 4} = \frac{\cancel{3}^1}{\cancel{3}_1} \times \frac{\cancel{2}^1}{\cancel{4}_2}$$

She has swum _____ km so far.

Do

1 Find the value of $\frac{1}{4} \times \frac{2}{3}$ by simplifying first.

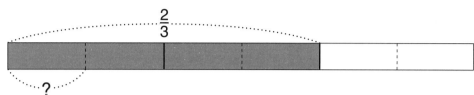

$$\frac{1}{4} \times \frac{2}{3} = \frac{1 \times \overset{1}{\cancel{2}}}{\underset{2}{\cancel{4}} \times 3}$$

$$= \frac{\square}{\square}$$

2 and 4 have a common factor of 2.

2 Find the value of $\frac{4}{5} \times \frac{5}{6}$ by simplifying first.

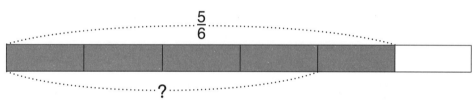

$$\frac{4}{5} \times \frac{5}{6} = \frac{\overset{2}{\cancel{4}} \times \overset{1}{\cancel{5}}}{\underset{1}{\cancel{5}} \times \underset{3}{\cancel{6}}}$$

$$= \frac{\square}{\square}$$

4 and 6 have a common factor of 2.
5 and 5 have a common factor of 5.

3 Find the value of $\frac{8}{9} \times \frac{3}{10}$.

$$\frac{8}{9} \times \frac{3}{10} = \frac{\cancel{8} \times \cancel{3}}{\cancel{9} \times \cancel{10}}$$

$$= \frac{\square}{\square}$$

The common factors of 8 and 10 are…
The common factors of 3 and 9 are…

5-7 Multiplying a Fraction by a Fraction — Part 2

4 Find the value of $\frac{5}{12} \times \frac{9}{10}$.

$$\frac{\cancel{5}^1}{\cancel{12}_4} \times \frac{\cancel{9}^3}{\cancel{10}_2} = \frac{}{}$$

5 Find the value by simplifying first.

(a) $\frac{1}{4} \times \frac{6}{7}$ (b) $\frac{4}{9} \times \frac{1}{12}$

(c) $\frac{3}{15} \times \frac{5}{8}$ (d) $\frac{4}{9} \times \frac{3}{8}$

(e) $\frac{3}{2} \times \frac{4}{9}$ (f) $\frac{5}{6} \times \frac{4}{15}$

(g) $\frac{9}{10} \times \frac{10}{9}$ (h) $\frac{15}{4} \times \frac{8}{3}$

6 Sharif ran $\frac{6}{10}$ km. Tyler ran $\frac{5}{4}$ as far as Sharif did. How many kilometers did Tyler run?

7 Write a word problem for $\frac{2}{3} \times \frac{9}{10}$.

Exercise 7 • page 120

Lesson 8
Multiplying Mixed Numbers

Think

A regular jump rope is $1\frac{1}{3}$ m long. A Double Dutch jump rope is $1\frac{1}{2}$ times as long as a regular jump rope. How many meters long is a Double Dutch jump rope?

Learn

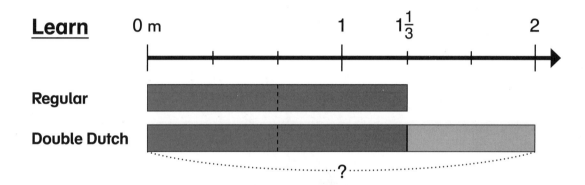

Express the mixed numbers as improper fractions.

$$1\frac{1}{2} \times 1\frac{1}{3} = \frac{\cancel{3}^1}{\cancel{2}_1} \times \frac{\cancel{4}^2}{\cancel{3}_1} = 2$$

A Double Dutch jump rope is _____ m long.

Do

1 Find the value of $3 \times 1\frac{1}{2}$.

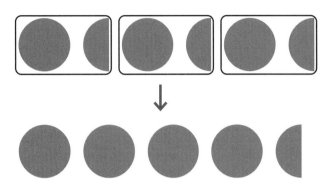

$3 \times 1\frac{1}{2} = 3 \times \frac{3}{2}$

$= \frac{3 \times 3}{2}$

$= \frac{\square}{\square}$

$= \square\frac{\square}{\square}$

2 Find the value of $\frac{1}{3} \times 1\frac{1}{2}$. Express the answer in simplest form.

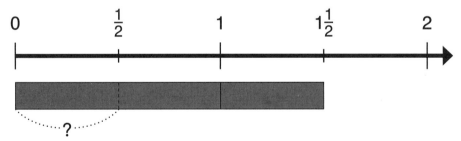

$\frac{1}{3} \times 1\frac{1}{2} = \frac{1}{3} \times \frac{3}{2}$

$= \frac{\square}{\square}$

3 Find the value of $1\frac{1}{6} \times \frac{3}{5}$. Express the answer in simplest form.

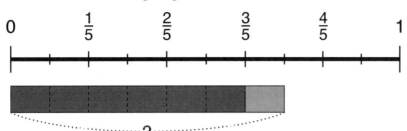

$1\frac{1}{6} \times \frac{3}{5} = \frac{7}{6} \times \frac{3}{5}$

$= \frac{\square}{\square}$

4 Find the product of $1\frac{3}{4}$ and $2\frac{1}{3}$.

$1\frac{3}{4} \times 2\frac{1}{3} = \frac{\square}{4} \times \frac{\square}{3} = \frac{\square}{\square} = \square\frac{\square}{\square}$

5 Find the values. Express each answer in simplest form.

(a) $7 \times 2\frac{5}{8}$

(b) $5\frac{1}{4} \times 6$

(c) $\frac{3}{5} \times 1\frac{1}{9}$

(d) $4\frac{2}{3} \times \frac{5}{7}$

(e) $1\frac{2}{5} \times \frac{7}{8}$

(f) $2\frac{1}{4} \times 1\frac{3}{5}$

6 The students in Class 5A ate $3\frac{1}{2}$ pizzas. The students in Class 5B ate $\frac{3}{4}$ as much pizza as Class 5A. How many pizzas did Class 5B eat?

7 Write a word problem for $3\frac{2}{3} \times \frac{3}{4}$.

Exercise 8 • page 123

Lesson 9
Word Problems — Part 2

Think

Yara had $300. She spent $\frac{2}{5}$ of it on soccer shoes and $\frac{1}{4}$ of the remainder on a soccer ball. How much money does she have left?

Learn

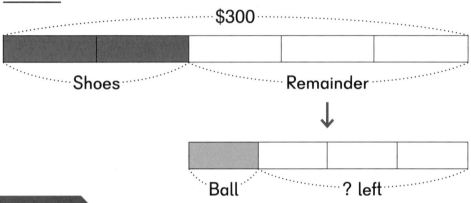

Method 1

I found the amount she spent on shoes first.

Spent on shoes: $\frac{2}{5} \times 300 = 120$

Remainder: $300 - 120 = 180$

Spent on ball: $\frac{1}{4} \times 180 = 45$

Money left: $180 - 45 = 135$

Method 2

$1 - \frac{2}{5} = \frac{3}{5}$

Remainder: $\frac{3}{5} \times 300 = 180$

Money left: $\frac{3}{4} \times 180 = 135$

I found the fraction that was the remainder first, and then the value of the remainder.

Method 3

5 larger units ⟶ 300

3 larger units ⟶ $\frac{1}{5} \times 300 \times 3 = 180$

4 smaller units ⟶ 180

3 smaller units ⟶ $\frac{1}{4} \times 180 \times 3 = 135$

I found the value of each of the two sizes of units.

Method 4

$\frac{3}{4} \times \frac{3}{5} \times 300 = \frac{9}{20} \times 300 = 135$

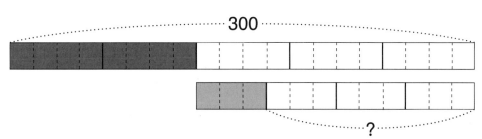

I found the fraction that she had left first. She has $\frac{3}{4}$ of $\frac{3}{5}$ left.

Yara has $ _____ left.

Do

1 Benjamin had $600. He spent $\frac{3}{4}$ of it on a bike. He spent $\frac{1}{3}$ of the money he had left on a helmet. How much did the helmet cost?

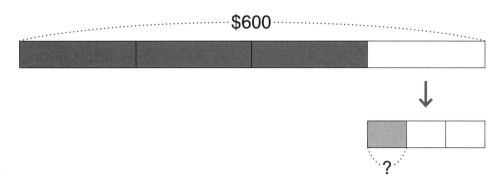

2 Santino baked 240 cookies. He sold $\frac{3}{4}$ of them and gave $\frac{3}{5}$ of the remainder to Ani. How many cookies did he give to Ani?

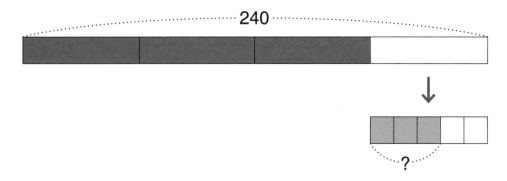

3 Asimah spent $\frac{1}{6}$ of her money on a shirt and $\frac{2}{3}$ of what was left on shoes. She then had $25 left. How much money did she have at first?

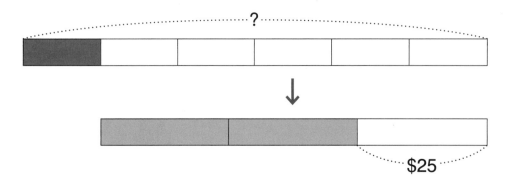

4. Sofia had some stickers. She used $\frac{2}{5}$ of them and gave $\frac{1}{2}$ of the remainder to Dion. Then, Dion gave $\frac{1}{3}$ of his stickers to Emma. Now Dion has 48 stickers left. How many stickers did Sofia have at first?

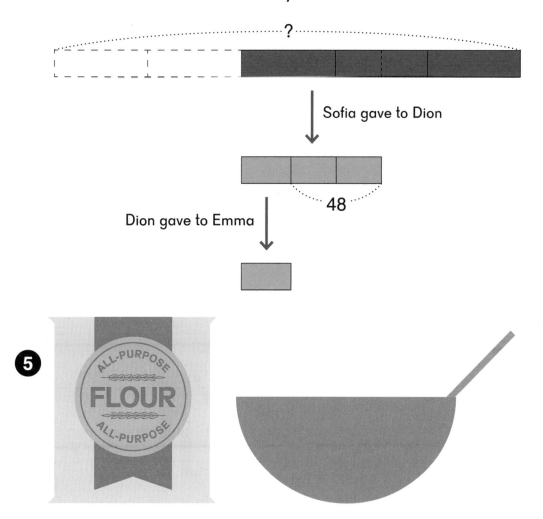

5. Jeremy had $1\frac{2}{5}$ kg of flour. He used $\frac{1}{3}$ of the flour to make bread and $\frac{1}{2}$ of the remaining flour to make muffins. How much flour did he use to make muffins?

6. Misha used $\frac{1}{2}$ of a roll of ribbon to wrap a present and $\frac{2}{3}$ of the rest of the roll to make a bow. If she had $\frac{3}{4}$ m of ribbon left, how long was the ribbon to start with?

Exercise 9 • page 126

Lesson 10
Fractions and Reciprocals

Think

Alex walks $\frac{3}{4}$ km to school every day. Emma walks $\frac{4}{3}$ times as far as Alex to school. How far does Emma walk to school?

Learn

Method 1

3 units ⟶ $\frac{3}{4}$

1 unit ⟶ $\frac{1}{3} \times \frac{3}{4} = \frac{1}{4}$

4 units ⟶ $4 \times \frac{1}{4} = 1$

$\frac{4}{3} > 1$ so Emma walks farther than Alex.

Method 2

$\frac{4}{3} \times \frac{3}{4} = \frac{\cancel{4} \times \cancel{3}}{\cancel{3} \times \cancel{4}} = 1$

$\frac{4}{3} \times \frac{3}{4} = \frac{12}{12} = 1$

Emma walks _____ km to school.

When the product of two numbers is 1, we say that one number is the **reciprocal** of the other number.

$\frac{3}{4}$ and $\frac{4}{3}$ are reciprocals. What other pairs of numbers that are reciprocals can you think of?

Do

① (a) What number is 4 times as much as $\frac{1}{4}$?

$4 \times \frac{1}{4} = \boxed{}$

$4 = \frac{4}{1}$ so the reciprocal of 4 is $\frac{1}{4}$.

(b) What number is $\frac{1}{5}$ of 5?

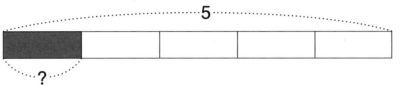

$\frac{1}{5} \times 5 = \boxed{}$

The reciprocal of $\frac{1}{5}$ is $\frac{5}{1}$.
$\frac{5}{1} = 5$

② (a) What number is $\frac{2}{3}$ of $\frac{3}{2}$?

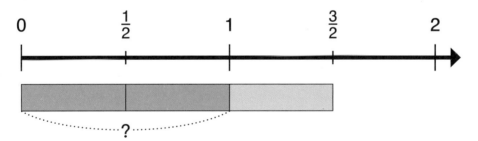

$\frac{2}{3} \times \frac{3}{2} = \boxed{}$

The reciprocal of $\frac{3}{2}$ is $\frac{2}{3}$.

(b) What is the reciprocal of $1\frac{1}{2}$?

$1\frac{1}{2} \times \boxed{} = 1$

$1\frac{1}{2} = \frac{3}{2}$ so the reciprocal is…

5-10 Fractions and Reciprocals

3 Find pairs of numbers that have a product of 1.

$\frac{4}{7}$ 2 $\frac{3}{8}$ $\frac{7}{4}$ $\frac{8}{5}$ $3\frac{2}{7}$ $\frac{8}{3}$ $\frac{5}{8}$ $\frac{7}{23}$ $\frac{1}{2}$

4 Find the reciprocal of each number.

(a) $\frac{8}{11}$

(b) $\frac{1}{5}$

(c) 6

(d) $\frac{9}{5}$

(e) $2\frac{1}{10}$

(f) $5\frac{3}{4}$

5 Find the missing numbers.

(a) $\frac{4}{9} \times \boxed{} = 1$

(b) $1\frac{1}{2} \times \boxed{} = 1$

(c) $\frac{1}{3} \times \frac{1}{2} \times \boxed{} = 1$

(d) $4 \times \boxed{} = 1$

(e) $\frac{1}{8} \times \boxed{} = 1$

(f) $\frac{3}{5} \times \frac{2}{7} \times \boxed{} = 1$

6 What is the reciprocal of 1?

7 Does 0 have a reciprocal? Explain why or why not.

Exercise 10 • page 130

Lesson 11
Practice B

1 Find the values. Express each answer in simplest form.

(a) $\frac{1}{3} \times \frac{5}{9}$

(b) $\frac{8}{5} \times \frac{1}{8}$

(c) $\frac{2}{5} \times \frac{5}{6}$

(d) $\frac{3}{4} \times \frac{8}{9}$

(e) $\frac{7}{8} \times \frac{3}{14}$

(f) $\frac{5}{3} \times \frac{2}{5}$

(g) $\frac{15}{4} \times \frac{8}{3}$

(h) $\frac{12}{5} \times \frac{10}{7}$

(i) $\frac{3}{4} \times 2\frac{2}{3}$

(j) $3\frac{4}{5} \times 4\frac{1}{2}$

(k) $\frac{2}{5} \times \frac{2}{3} \times 5 \times 3$

(l) $\frac{3}{8} \times \frac{2}{3} \times 720$

2 Find the values. Express each answer in simplest form.

(a) $6 - 15 \times \frac{1}{3} \times \frac{5}{9}$

(b) $\frac{1}{2} \times (11 + 25) \times \frac{1}{4}$

(c) $(\frac{1}{3} \times 2) + (\frac{1}{3} \times 19)$

(d) $40 \div (\frac{1}{3} \times 18)$

3 Find the reciprocal of each number.

(a) 7

(b) $2\frac{1}{2}$

(c) $\frac{7}{10}$

(d) $4\frac{2}{3}$

4 Justin earned some money cutting lawns. He saved $\frac{3}{4}$ of it and gave the rest equally to 5 charities. Each charity received $175. How much money did Justin earn cutting lawns?

5 A baker had some flour in a container. After using 12 lb to bake pita bread and $\frac{1}{3}$ of the remainder to bake naan bread, he had 12 lb left. How much flour was in the container?

6 A fruit vendor sold $\frac{2}{3}$ of his apples in the morning and $\frac{1}{6}$ of his remaining apples in the afternoon. He sold 195 apples altogether. How many apples did he have left?

7 Michelle spent $\frac{2}{3}$ of her money on books and $\frac{1}{4}$ of the remainder on school supplies. She had $144 left. How much money did she have at first?

8 Kalama's mother made $2\frac{3}{5}$ L of lemonade. Kalama drank $\frac{1}{2}$ of the lemonade. How many liters of lemonade were left?

9 Rope A is $5\frac{3}{5}$ m long. Rope B is $1\frac{2}{3}$ times as long as Rope A. Rope C is $1\frac{3}{7}$ times as long as Rope A. How long are the three ropes altogether in meters?

10 Tomas had a board that was $3\frac{1}{5}$ m long. He sawed off $1\frac{3}{5}$ m of the board and used $\frac{1}{3}$ of the remaining board for a ramp. How long was the ramp in meters?

Exercise 11 • page 132

Chapter 6

Division of Fractions

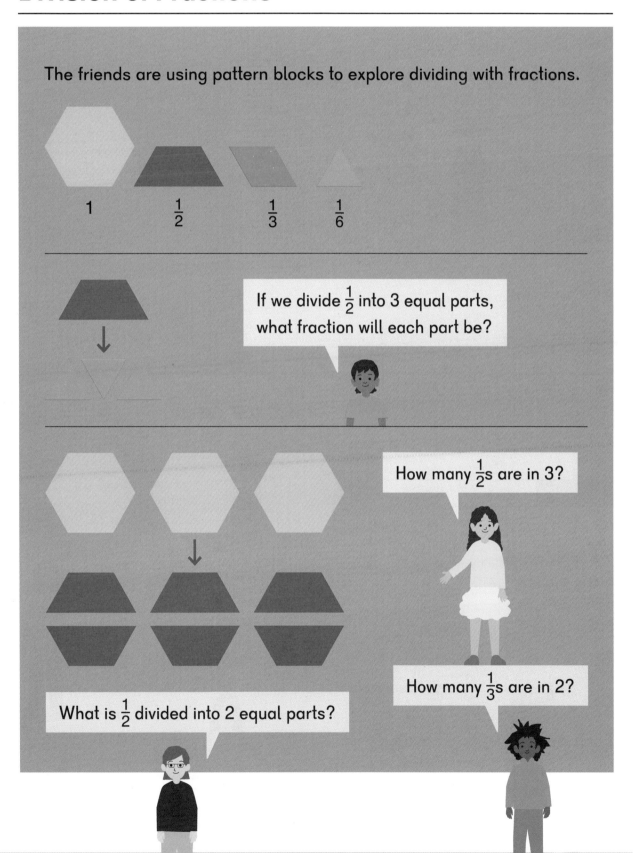

Lesson 1
Dividing a Unit Fraction by a Whole Number

Think

A $\frac{1}{2}$ m long string is cut into 3 equal length pieces. How long is each piece in meters?

$3 \times ? = \frac{1}{2}$

Learn

$\frac{1}{2} \div 3$

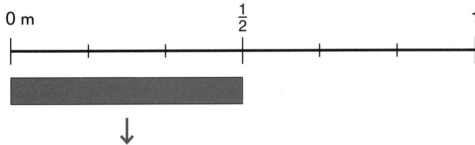

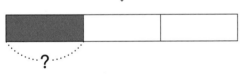

$\frac{1}{2} \div 3 = \frac{1}{3}$ of $\frac{1}{2}$

$= \frac{1}{3} \times \frac{1}{2}$

$= \frac{1}{6}$

We can also write $\frac{1}{2} \div 3 = \frac{1}{2} \times \frac{1}{3}$.

Dividing by 3 is the same as multiplying by $\frac{1}{3}$. $\frac{1}{3}$ is the reciprocal of 3.

Each piece is _____ m long.

Do

1 Divide $\frac{1}{4}$ by 2.

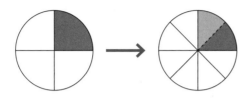

Dividing by 2 is the same as multiplying by $\frac{1}{2}$.

$\frac{1}{4} \div 2 = \frac{1}{4} \times \frac{1}{2}$

$= \frac{}{}$

2 Divide $\frac{1}{5}$ by 4.

What is the reciprocal of 4?

$\frac{1}{5} \div 4 = \frac{1}{5} \times \frac{}{}$

$= \frac{}{}$

3 Divide $\frac{1}{3}$ by 5.

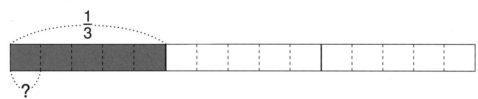

$\frac{1}{3} \div 5 = \frac{}{} \times \frac{}{}$

$= \frac{}{}$

What is $\frac{1}{5}$ of $\frac{1}{3}$?

$5 \times ? = \frac{1}{3}$

4 Divide $\frac{1}{2}$ by 1.

$\frac{1}{2} \div 1 = \boxed{}$

What is the value of any number divided by 1?
What is the reciprocal of 1?

5 Divide.

(a) $\frac{1}{3} \div 7$ (b) $\frac{1}{7} \div 3$

(c) $\frac{1}{6} \div 2$ (d) $\frac{1}{2} \div 6$

(e) $\frac{1}{8} \div 5$ (f) $\frac{1}{9} \div 3$

6 (a) $\frac{1}{5} \div \boxed{} = \frac{1}{10}$

(b) $\frac{1}{4} \div \boxed{} = \frac{1}{20}$

(c) $\frac{1}{3} \div \boxed{} = \frac{1}{6}$

7 $\frac{1}{5}$ L of juice is poured equally into 5 cups. How many liters of juice are in each cup?

8 Write a word problem for $\frac{1}{3} \div 2$.

Exercise 1 • page 137

Lesson 2
Dividing a Fraction by a Whole Number

Think

Dion and Mei are sharing $\frac{4}{5}$ of a cake equally. What fraction of the cake will each person get?

Since we are dividing $\frac{4}{5}$ by 2, the answer will be less than $\frac{4}{5}$.

Learn

$\frac{4}{5} \div 2$

Method 1

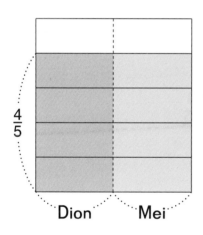

Dion · Mei

$\frac{4}{5} \div 2 = \frac{1}{2}$ of $\frac{4}{5}$

$= \frac{1}{2} \times \frac{4}{5}$

$= \frac{4}{10}$

$= \frac{2}{5}$

Dividing by 2 is the same as multiplying by its reciprocal, $\frac{1}{2}$.

Method 2

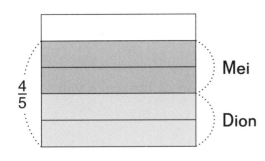

I thought of the fifths as units. 4 fifths ÷ 2 = 2 fifths

$\frac{4}{5} \div 2 = \frac{2}{5}$

Method 3

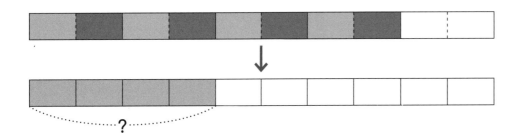

$\frac{1}{5} \div 2 = \frac{1}{10}$

$\frac{4}{5} \div 2 = 4 \times \frac{1}{10}$

$\phantom{\frac{4}{5} \div 2} = \frac{4}{10}$

$\phantom{\frac{4}{5} \div 2} = \frac{2}{5}$

I know that $\frac{1}{5} \div 2 = \frac{1}{10}$. $\frac{4}{5} \div 2$ is just 4 times that.

Each person will get _____ of the cake.

Do

1.  →

   Divide $\frac{3}{4}$ by 3.

   3 fourths ÷ 3 = ? fourth

   $\frac{3}{4} \div 3 = \frac{\square}{\square}$

2. Divide $\frac{6}{7}$ by 2.

   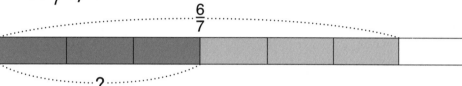

   $\frac{6}{7} \div 2 = \frac{\square}{\square}$

   6 sevenths ÷ 2 = ? sevenths

3. Divide $\frac{2}{3}$ by 4. Express the answer in simplest form.

   (a) $\frac{2}{3} \div 4 = \frac{2}{3} \times \frac{1}{4}$

   $= \square$

   $\frac{1}{4}$ of $\frac{2}{3}$ →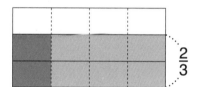

   (b) $\frac{2}{3} \div 4 = \frac{2 \times 2}{3 \times 2} \div 4$

   $= \frac{4}{6} \div 4$

   $= \square$

   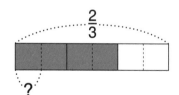

    2 is not divisible by 4 so express $\frac{2}{3}$ as $\frac{4}{6}$. 4 sixths ÷ 4 = ? sixths

6-2 Dividing a Fraction by a Whole Number

4) A 3-m pipe weighs $\frac{9}{10}$ kg. How many kilograms does 1 m of the pipe weigh?

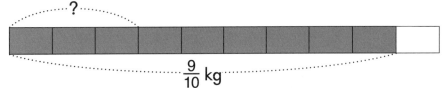

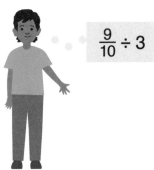

$\frac{9}{10} \div 3$

5) Find the values.

(a) $\frac{4}{7} \div 4$ (b) $\frac{7}{4} \div 4$ (c) $\frac{8}{9} \div 4$

(d) $\frac{3}{5} \div 6$ (e) $\frac{2}{3} \div 10$ (f) $\frac{5}{8} \div 2$

6) Dion wants to cut $\frac{2}{3}$ m of rope into 6 pieces of equal length. How long in meters should he cut each piece?

7) $\frac{9}{10}$ L of juice was poured into 6 cups. How many liters of juice are in each cup?

8) Write a word problem for $\frac{4}{5} \div 2$.

Exercise 2 • page 139

Lesson 3
Practice A

1 Find the values. Express each answer in simplest form.

(a) $\frac{1}{5} \div 3$

(b) $\frac{1}{3} \div 8$

(c) $\frac{3}{5} \div 7$

(d) $\frac{5}{6} \div 5$

(e) $\frac{2}{9} \div 2$

(f) $\frac{9}{10} \div 3$

(g) $\frac{3}{4} \div 6$

(h) $\frac{9}{8} \div 1$

(i) $\frac{3}{4} \div 2 \times 3$

(j) $\frac{1}{2} \div 2 \div 3$

(k) $(\frac{1}{2} + \frac{1}{3}) \div 5$

(l) $(\frac{5}{8} - \frac{1}{8}) \div 6$

2 A ribbon that is $\frac{3}{5}$ m long is cut into 6 pieces of equal length. How long is each piece?

3 5 identical boxes of paper clips weigh $\frac{2}{3}$ kg altogether. How many kilograms does 1 box of paper clips weigh?

4 $\frac{3}{4}$ lb of almonds are put equally into 6 bags. What is the weight of the almonds in each bag?

5 The perimeter of a square is $\frac{2}{5}$ m. What is the length of one side of the square in meters?

Exercise 3 • page 141

Lesson 4
Dividing a Whole Number by a Unit Fraction

Think

Mei has 4 m of ribbon. She wants to cut smaller pieces so that each piece is $\frac{1}{3}$ m long. How many pieces of ribbon will she have?

How many $\frac{1}{3}$ m pieces are in 1 m?

Learn

$4 \div \frac{1}{3}$

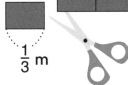

$1 \div \frac{1}{3} = 3$

$4 \div \frac{1}{3} = 4 \times 3$

$\qquad = 12$

Dividing by $\frac{1}{3}$ is the same as multiplying by its reciprocal, 3.

$? \times \frac{1}{3} = 4$
$? = 4 \div \frac{1}{3}$
$? = 4 \times 3$

Mei will have _____ pieces of ribbon.

Do

1 (a) If we cut 1 pie into pieces that are $\frac{1}{4}$ pie each, how many pieces will we have?

How many $\frac{1}{4}$s are in 1 whole?

$1 \div \frac{1}{4} = \boxed{}$

(b) If we cut 3 pies into pieces that are $\frac{1}{4}$ pie each, how many pieces will we have?

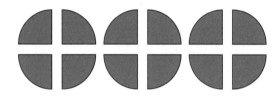

How many $\frac{1}{4}$s are in 3 wholes?

$3 \div \frac{1}{4} = \boxed{}$

2 A road crew is paving a road that is 5 miles long. They can pave $\frac{1}{3}$ mi in one day. How many days will it take them to pave the road?

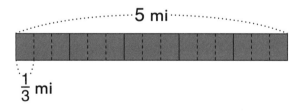

How many $\frac{1}{3}$s are in 5?

Days it takes to pave 1 mile ⟶ $1 \div \frac{1}{3} = 3$

Days it takes to pave 5 miles ⟶ $5 \div \frac{1}{3} = 5 \times \boxed{}$

$= \boxed{}$

6-4 Dividing a Whole Number by a Unit Fraction

3. Sofia has 4 L of water. She puts $\frac{1}{8}$ L of water in each glass. How many glasses will she need?

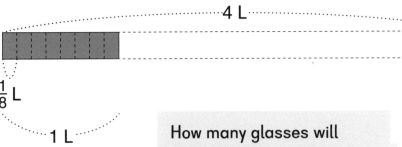

$4 \div \frac{1}{8} = 4 \times \square$

$= \square$

How many glasses will Sofia need for 1 L of water?

4. $\frac{1}{4}$ lb of coffee costs $2. How much does 1 lb of coffee cost?

$\frac{1}{4}$ lb ⟶ $2

1 lb ⟶ $2 $\div \frac{1}{4}$ = $2 $\times \square$ = $\square$

5. Find the values.

(a) $1 \div \frac{1}{8}$ (b) $2 \div \frac{1}{3}$

(c) $9 \div \frac{1}{2}$ (d) $6 \div \frac{1}{5}$

(e) $12 \div \frac{1}{9}$ (f) $10 \div \frac{1}{10}$

6. Write a word problem for $3 \div \frac{1}{2}$.

Exercise 4 • page 143

Lesson 5
Dividing a Whole Number by a Fraction

Think

Dion has some floral tape that is 4 m long. He wants to cut it into shorter pieces that are each $\frac{2}{3}$ m long. How many shorter pieces can he make?

$? \times \frac{2}{3} = 4$

Learn

$4 \div \frac{2}{3}$

Method 1

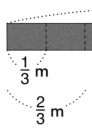

$4 \div \frac{1}{3} = 4 \times 3$

$4 \div \frac{2}{3} = \frac{4 \times 3}{2}$

$= \frac{12}{2}$

$= 6$

There are 12 thirds in 4. How many groups of 2 thirds are there? 12 thirds ÷ 2 thirds = 6

$\frac{4 \times 3}{2} = 4 \times \frac{3}{2}$

We are multiplying 4 by the reciprocal of $\frac{2}{3}$.

Method 2

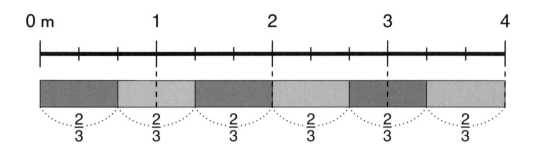

In 1 m there are $1\frac{1}{2} = \frac{3}{2}$ two-thirds meter pieces.

In 4 m there are $4 \times \frac{3}{2}$ two-thirds meter pieces.

$4 \div \frac{2}{3} = 4 \times \frac{3}{2}$ The reciprocal of $\frac{2}{3}$ is...

$= \frac{12}{2}$

$= 6$

He can make _____ shorter pieces.

Do

❶ How many $\frac{2}{3}$s are in 2?

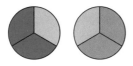

$2 \div \frac{2}{3} = \boxed{}$

❷ There are 3 pies. If we put $\frac{3}{5}$ of a pie on each plate, how many plates will we need?

$3 \div \frac{1}{5} = 3 \times 5$

$3 \div \frac{3}{5} = \frac{3 \times 5}{3}$

$= \boxed{}$

In 3 pies, there are 15 pieces that are $\frac{1}{5}$ of a pie.

❸ It takes one can of paint to paint $\frac{3}{4}$ of a wall. How many cans of paint will it take to paint 4 of the same sized walls?

$4 \div \frac{3}{4} = 4 \times \frac{4}{3}$

$= \frac{\boxed{}}{3}$

$= \boxed{} \frac{\boxed{}}{3}$

1 can ⟶ $\frac{3}{4}$ wall

$\frac{1}{3}$ can ⟶ $\frac{1}{4}$ wall

$\frac{4}{3}$ can ⟶ 1 wall

6-5 Dividing a Whole Number by a Fraction

4. Emma bikes 5 miles, which is $\frac{2}{3}$ of the way along the trail. How long is the trail in miles?

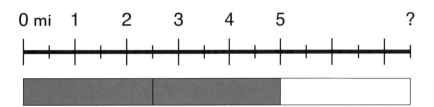

$5 \div \frac{2}{3} = 5 \times \frac{\square}{2}$

$= \frac{\square}{2}$

$= \square \frac{\square}{\square}$

$\frac{2}{3}$ of the way ⟶ 5 mi

$\frac{1}{3}$ of the way ⟶ $\frac{5}{2}$ mi

1 whole way ⟶ $3 \times \frac{5}{2}$

5 miles is $\frac{2}{3}$ of ?

5. Find the values. Express each answer in simplest form.

(a) $3 \div \frac{3}{4}$
(b) $4 \div \frac{4}{5}$
(c) $6 \div \frac{2}{7}$

(d) $3 \div \frac{2}{3}$
(e) $7 \div \frac{3}{4}$
(f) $5 \div \frac{3}{8}$

(g) $3 \div \frac{3}{2}$
(h) $7 \div \frac{4}{3}$
(i) $5 \div \frac{8}{3}$

6. A store manager has 8 lb of coffee. She wants to put $\frac{2}{3}$ lb of coffee in each bag. How many bags will she need?

7. Write a word problem for $6 \div \frac{3}{4}$.

Exercise 5 • page 146

Lesson 6
Word Problems

Think

Alex had 6 m of ribbon. He used $\frac{2}{3}$ of it to wrap presents and the rest of it to make bows. He used $\frac{1}{2}$ m of ribbon for each bow. How many bows did he make?

Learn

Method 1

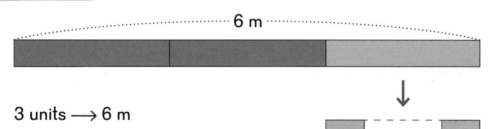

3 units ⟶ 6 m

1 unit ⟶ 6 m ÷ 3 = 2 m

Number of $\frac{1}{2}$ m pieces in 2 m: 2 m ÷ $\frac{1}{2}$ m = 4

Method 2

$1 - \frac{2}{3} = \frac{1}{3}$

$\frac{1}{3} \times 6$ m = 2 m

2 m ÷ $\frac{1}{2}$ m = 4

He made _____ bows.

Do

① A dairy farm produced 60 L of milk. They used $\frac{1}{2}$ of it to make cheese and put $\frac{1}{2}$ of the remainder into bottles that hold $\frac{3}{4}$ L each. How many bottles of milk are there?

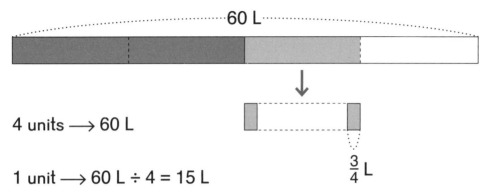

4 units ⟶ 60 L

1 unit ⟶ 60 L ÷ 4 = 15 L

Number of bottles: 15 L ÷ $\frac{3}{4}$ L = ☐

How many $\frac{3}{4}$ L are in 15 L?

② A box containing 10 bags of peanuts weighs $\frac{3}{4}$ lb. The box itself weighs $\frac{1}{8}$ lb. What is the weight of each bag of peanuts?

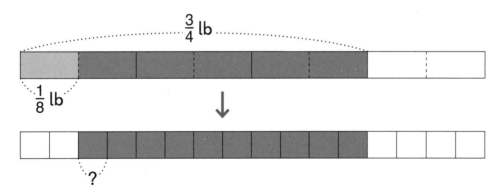

Weight of peanuts: $\frac{3}{4}$ lb − $\frac{1}{8}$ lb = ☐ lb $(\frac{3}{4} - \frac{1}{8}) \div 10$

Weight of 1 bag of peanuts: ☐ lb ÷ 10 = ☐ lb

3 A baker put $\frac{4}{5}$ kg of flour equally in 2 containers, A and B. He used $\frac{1}{10}$ kg from Container B to bake bread. How much flour is left in Container B?

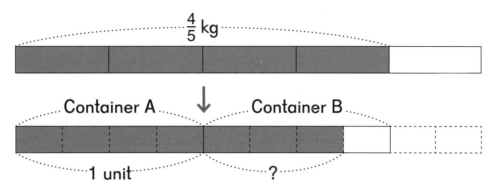

Flour put in each container ⟶ $\frac{4}{5}$ kg ÷ 2 = ☐ kg

$\frac{4}{5} \div 2 - \frac{1}{10}$

Flour left in Container B ⟶ ☐ kg − $\frac{1}{10}$ kg = ☐ kg

4 A florist spent $18 on $\frac{2}{3}$ m of ribbon and some flower pots. The flower pots cost $12. How much would 1 m of ribbon cost?

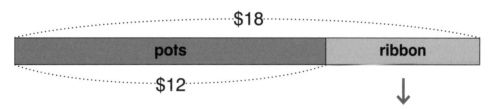

(a) $\frac{2}{3}$ m ⟶ $18 − $12 = $6

$\frac{1}{3}$ m ⟶ $6 ÷ 2 = $☐

1 m ⟶ $☐ × 3 = $☐

(b) ($18 − $12) ÷ $\frac{2}{3}$ = $☐

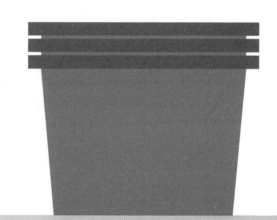

Exercise 6 • page 149

Lesson 7
Practice B

1 Find the values. Express each answer in simplest form.

(a) $4 \div \frac{1}{3}$

(b) $10 \div \frac{1}{6}$

(c) $6 \div \frac{6}{7}$

(d) $2 \div \frac{2}{5}$

(e) $5 \div \frac{3}{5}$

(f) $12 \div \frac{3}{4}$

(g) $9 \div \frac{5}{7}$

(h) $4 \times 2 \div \frac{1}{2}$

(i) $120 - 15 \div \frac{3}{5}$

(j) $4 \times 3 \div \frac{3}{4}$

(k) $8 \div (\frac{3}{4} - \frac{1}{2})$

(l) $(4\frac{1}{3} + \frac{4}{6}) \div \frac{5}{6}$

2 A cook has 3 lb of pasta. Each pasta dish needs $\frac{1}{8}$ lb of pasta. How many pasta dishes can he make?

3 A bullet train can travel 40 mi in $\frac{1}{5}$ h. How far can the train travel in 1 h?

4 A $\frac{1}{4}$ m long rope was cut into 3 equal pieces. How long is each piece of rope in meters?

5 Mary Jane had $\frac{3}{5}$ kg of apples. She ate $\frac{1}{10}$ kg and used the rest to make 5 apple pies. How many kilograms of apples did she use for each pie?

6 A store employee had 75 lb of light roast coffee and 60 lb of dark roast coffee. She made bags of $\frac{3}{4}$ lb of coffee to sell. How many bags of each type of coffee did she make?

Exercise 7 • page 152

Chapter 7

Measurement

Lesson 1
Fractions and Measurement Conversions

Think

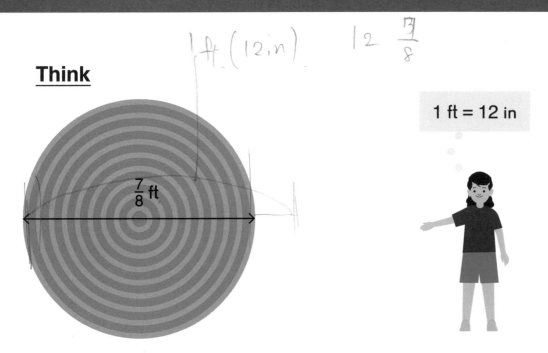

The diameter of a frisbee is $\frac{7}{8}$ ft. What is the diameter of the frisbee in inches?

Learn

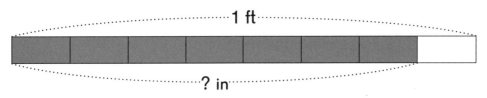

1 ft = 12 in

$\frac{7}{8}$ ft = $\frac{7}{8}$ × 12 in

$= \frac{7 \times \overset{3}{\cancel{12}}}{\underset{2}{\cancel{8}}}$ in

$= \frac{21}{2}$ in

$= 10\frac{1}{2}$ in

Convert 1 foot to inches and multiply by $\frac{7}{8}$.

The diameter of the frisbee is _____ in.

Customary Conversions		Metric Conversions	
1 ft........12 in	1 lb........16 oz	1 km....1,000 m	1 L....1,000 mL
1 yd........3 ft	1 day........24 h	1 m........100 cm	1 kg......1,000 g
1 qt........4 c	1 min........60 s	1 cm........10 mm	
1 gal........4 qt	1 h........60 min		

Do

1 Mei played badminton for $\frac{3}{4}$ of an hour. How many minutes did she play badminton?

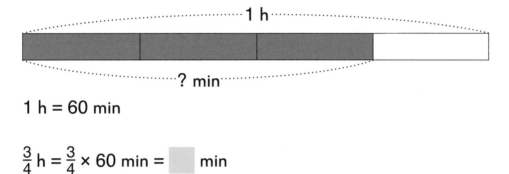

1 h = 60 min

$\frac{3}{4}$ h = $\frac{3}{4}$ × 60 min = ☐ min

2 A jug holds 1 gal of water. It is $\frac{4}{5}$ full. How many quarts of water are in the jug? Express the answer as a mixed number in simplest form.

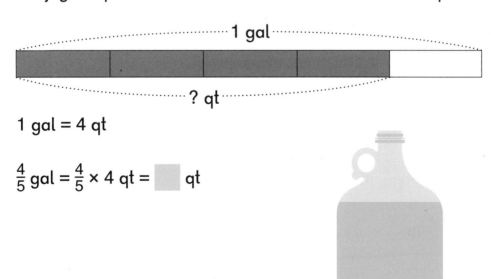

1 gal = 4 qt

$\frac{4}{5}$ gal = $\frac{4}{5}$ × 4 qt = ☐ qt

7-1 Fractions and Measurement Conversions

3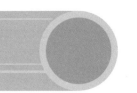

A metal pipe weighs $2\frac{3}{4}$ pounds. How many ounces does the pipe weigh?

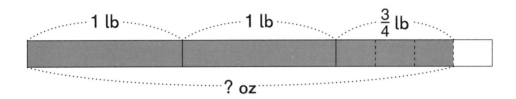

1 lb = 16 oz

2 lb = 2 × 16 oz = 32 oz

$\frac{3}{4}$ lb = $\frac{3}{4}$ × 16 oz = ☐ oz

$2\frac{3}{4}$ lb = 32 oz + ☐ oz = ☐ oz

4 A tennis racket is $2\frac{1}{3}$ ft long. How long is the tennis racket in feet and inches?

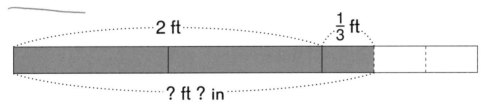

1 ft = 12 in

$\frac{1}{3}$ ft = $\frac{1}{3}$ × 12 in = ☐ in

$2\frac{1}{3}$ ft = 2 ft ☐ in

5 An exercise ball weighs $3\frac{1}{2}$ kg. How much does it weigh in grams?

$3\frac{1}{2}$ kg = 3 kg + $\frac{1}{2}$ kg = ? g + ? g

1 kg = 1,000 g

3 kg = 3 × 1,000 g = ☐ g

$\frac{1}{2}$ kg = $\frac{1}{2}$ × 1,000 g = ☐ g

$3\frac{1}{2}$ kg = ☐ g

6 (a) $\frac{4}{5}$ km = ☐ m (b) $\frac{3}{4}$ m = ☐ cm

(c) $\frac{1}{2}$ cm = ☐ mm (d) $\frac{3}{4}$ lb = ☐ oz

(e) $\frac{3}{8}$ ft = ☐ in (f) $\frac{2}{3}$ h = ☐ min

7 (a) $2\frac{1}{2}$ min = ☐ min ☐ s (b) $2\frac{1}{2}$ min = ☐ s

(c) $3\frac{1}{2}$ gal = ☐ gal ☐ qt (d) $3\frac{1}{2}$ gal = ☐ qt

(e) $2\frac{1}{2}$ kg = ☐ kg ☐ g (f) $2\frac{1}{2}$ kg = ☐ g

(g) $8\frac{7}{10}$ cm = ☐ cm ☐ mm (h) $8\frac{7}{10}$ cm = ☐ mm

(i) $8\frac{7}{10}$ m = ☐ m ☐ cm (j) $8\frac{7}{10}$ m = ☐ cm

Exercise 1 • page 157

Lesson 2
Fractions and Area

Think

A $\frac{3}{4}$ m by $\frac{1}{3}$ m rectangular section of a 1 square meter poster board is painted yellow. What is the area of the section that is painted yellow?

Learn

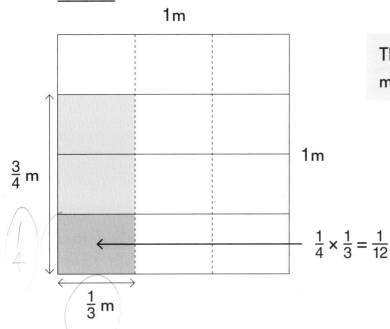

There are 3 one-twelfth square meter units in the yellow section.

$\frac{1}{4} \times \frac{1}{3} = \frac{1}{12}$

$\frac{3}{4} \times \frac{1}{3} = \frac{3}{12} = \frac{1}{4}$

The area of the section that is painted yellow is _____ m².

We can find the area of a rectangle by multiplying the length by the width even when the side lengths are fractions.

Do

1 Find the area of the large square and the area of the smaller yellow square.

$1 \times 1 =$ ☐

Area of large square = ☐ in²

$\frac{1}{2} \times \frac{1}{2} =$ ☐

Area of small square = ☐ in²

The area of the small square is $\frac{1}{4}$ of the area of the large square.

2 Find the area of the largest rectangle.

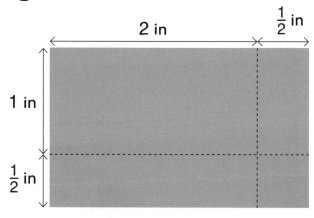

$2\frac{1}{2} \times 1\frac{1}{2} = \frac{5}{2} \times \frac{3}{2}$

I can find the area of each smaller rectangle first.

7-2 Fractions and Area

3. Find the area of each rectangle.

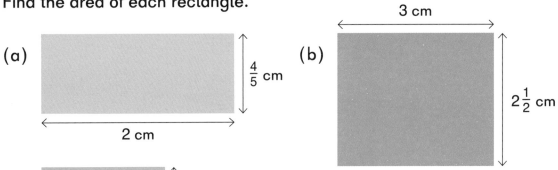

(a) 2 cm × 4/5 cm

(b) 3 cm × 2 1/2 cm

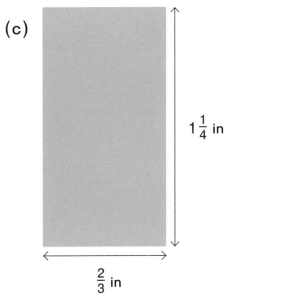

(c) 2/3 in × 1 1/4 in

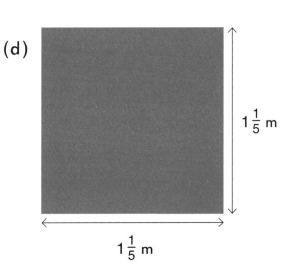

(d) 1 1/5 m × 1 1/5 m

4. The figures below are composed of rectangles. Find the area of each figure.

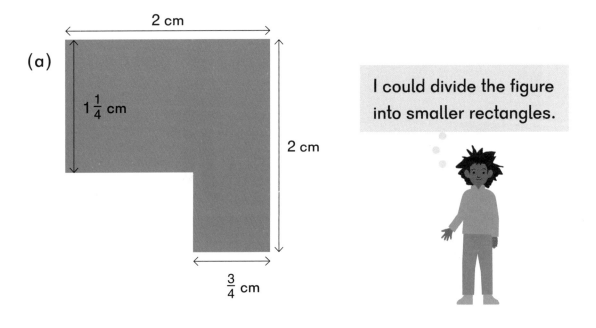

(a)

I could divide the figure into smaller rectangles.

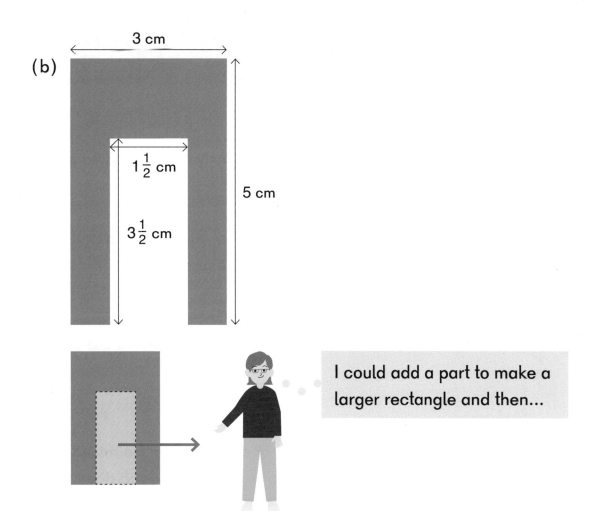

5 There is a grassy area around a rectangular garden. How many square meters of grass are there?

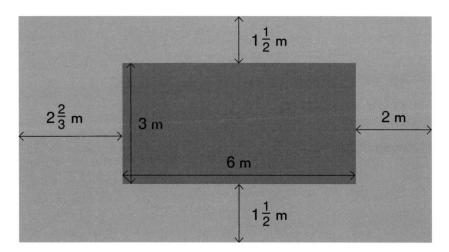

Exercise 2 • page 160

Lesson 3
Practice A

1. Jason played tennis for $\frac{2}{3}$ of an hour. How many minutes did he play tennis?

2. Maria needs $\frac{3}{8}$ lb of flour for a muffin recipe. She wants to double the recipe. How many ounces of flour does she need?

3. Last week Jackie ran $\frac{4}{5}$ km every day for 6 days. How far did she run in meters?

4. Danny had $\frac{3}{4}$ qt of apple juice and 2 qt of lemonade. He mixed $\frac{2}{3}$ of each juice to make fruit punch. How many cups of fruit punch does he have?

5. Find the area of each figure.

(a)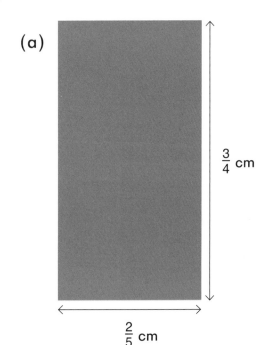

(b)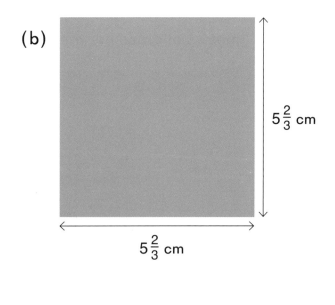

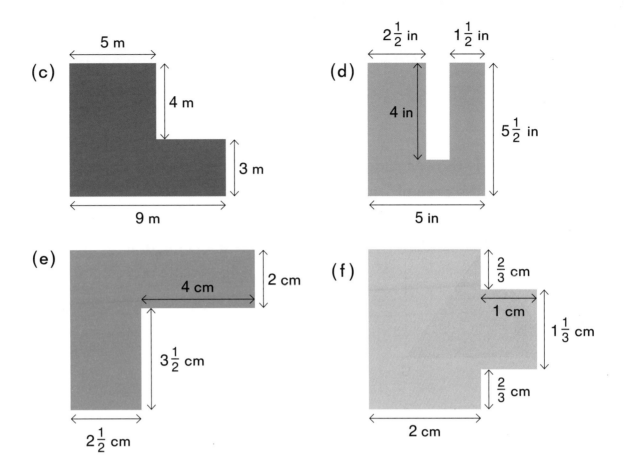

6. There is a grassy area around a rectangular pond.

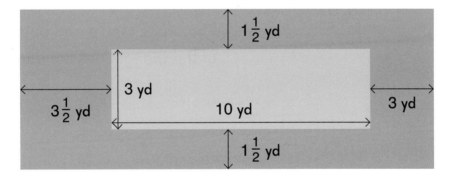

(a) How many square yards of grass are there?

(b) How many square feet are in one square yard?

(c) How many square feet of grass are there?

Exercise 3 • page 163

Lesson 4
Area of a Triangle — Part 1

Think

Triangle ABC is drawn inside a rectangle with a length of 6 cm and a width of 4 cm. Find the area of Triangle ABC.

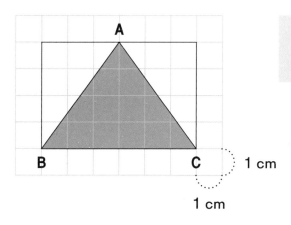

How can we use the area of the rectangle to find the area of the triangle?

Learn

Method 1

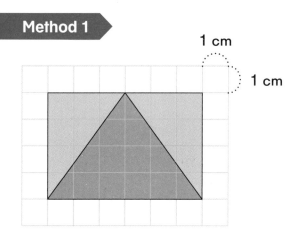

The areas inside and outside of the triangle are the same.

Area of Triangle ABC = $\frac{1}{2} \times (6 \times 4)$ = 12 cm²

Area of Triangle ABC = $\frac{1}{2} \times$ (Length × Width)

Method 2

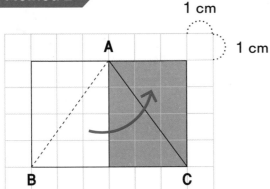

Area of Triangle ABC = ($\frac{1}{2}$ × 6) × 4 = 12 cm²

Area of Triangle ABC = ($\frac{1}{2}$ × Length) × Width

Method 3

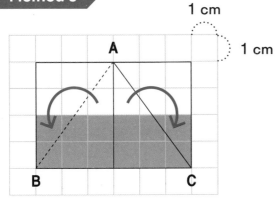

I cut the triangle halfway along the width of the rectangle, and moved the pieces to make the smaller rectangle.

Area of Triangle ABC = ($\frac{1}{2}$ × 4) × 6 = 12 cm²

Area of Triangle ABC = ($\frac{1}{2}$ × Width) × Length

The area of Triangle ABC is _____ cm².

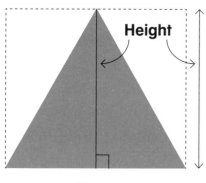

The area of the triangle is half the area of a rectangle that has a length and width the same as the triangle's base and height.

Any side of a triangle can be considered as the base.

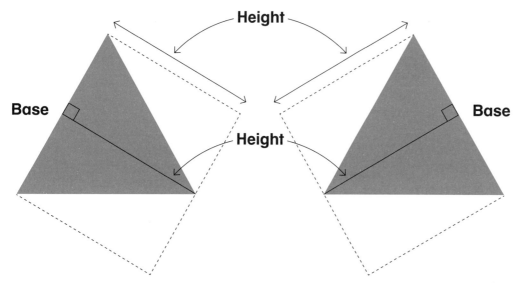

On a triangle one side is called the **base**. The **height** is the perpendicular distance from the base to the opposite vertex.

Area of Triangle = $\frac{1}{2}$ × Base × Height

I wonder if this is true for other kinds of triangles?

Do

1 Find the area of each triangle. Each square represents 1 square centimeter.

(a)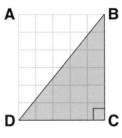

Area of Rectangle ABCD = 5 × 6 = ☐ cm²

Area of Triangle BCD = $\frac{1}{2}$ × 5 × 6 = ☐ cm²

(b)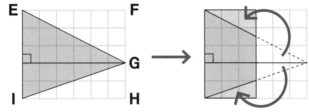

Area of Rectangle EFHI = ☐ × ☐ = ☐ cm²

Area of Triangle EGI = $\frac{1}{2}$ × ☐ × ☐ = ☐ cm²

(c)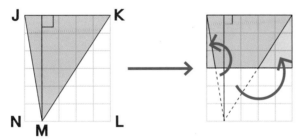

Area of Rectangle JKLN = ☐ × ☐ = ☐ cm²

Area of Triangle JKM = $\frac{1}{2}$ × ☐ × ☐ = ☐ cm²

Why do all three triangles have the same area?

7-4 Area of a Triangle — Part 1

2. Find the area of the triangle. Each square represents 1 square centimeter.

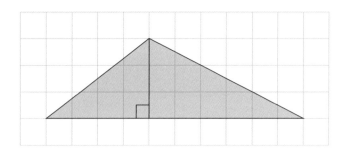

3. Identify a base and the corresponding height for each triangle.

(a) Base = DF

Height =

(b) Base =

Height =

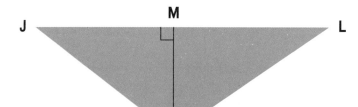

(c) Base =

Height =

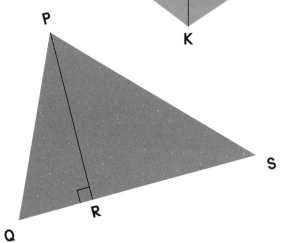

4 Find the area of each triangle.

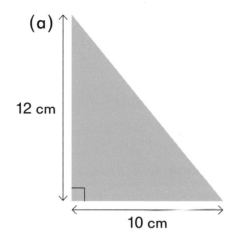

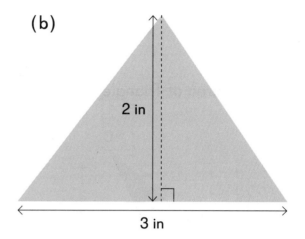

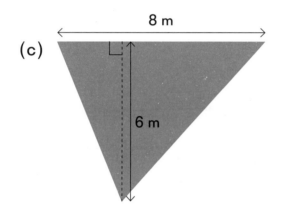

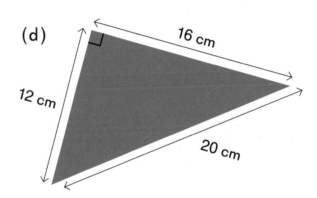

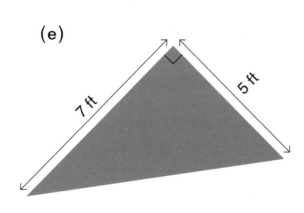

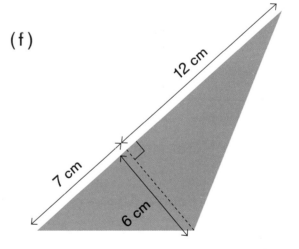

Exercise 4 • page 166

Lesson 5
Area of a Triangle — Part 2

Think

Find the area of Triangle ABC.

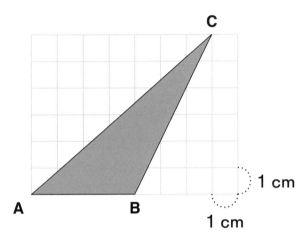

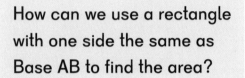

How can we use a rectangle with one side the same as Base AB to find the area?

Learn

| Method 1 |

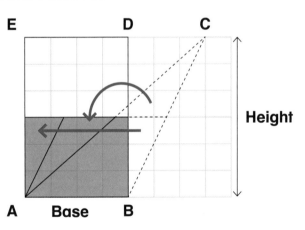

The height of the triangle is perpendicular to the base. The height of the green rectangle is half of the height of the triangle.

Area of Rectangle ABDE = 4 × 6 = 24 cm²

Area of Triangle ABC = $\frac{1}{2}$ of the area of Rectangle ABDE

$= \frac{1}{2} \times 4 \times 6 = 12$ cm²

$= \frac{1}{2} \times$ base × height

Method 2

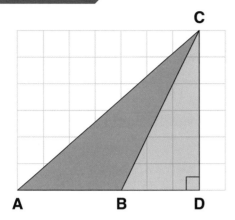

We could instead make a larger triangle ACD and subtract the area of Triangle BCD.

Area of Triangle ACD = $\frac{1}{2} \times (4 + 3) \times 6 = \frac{1}{2} \times 42 = 21$ cm²

Area of Triangle BCD = $\frac{1}{2} \times 3 \times 6 = \frac{1}{2} \times 18 = 9$ cm²

Area of Triangle ABC = 21 − 9 = 12 cm²

The differences between the bases of triangles ACD and BCD is 4.
$\frac{1}{2} \times (7 - 3) \times 6 = \frac{1}{2} \times 4 \times 6 = \frac{1}{2} \times$ base $\times$ height

We can use the formula $\frac{1}{2} \times$ base $\times$ height even when the height is outside of the base.

The area of Triangle ABC is _____ cm².

Do

1 Each square represents 1 square centimeter.

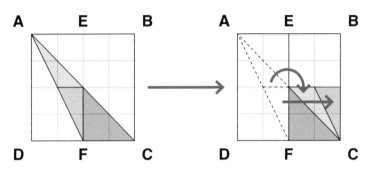

Area of Rectangle FEBC = ☐ × ☐ = ☐ cm²

Area of Triangle ACF = $\frac{1}{2}$ × ☐ × ☐ = ☐ cm²

2 Find the area of each triangle.

(a)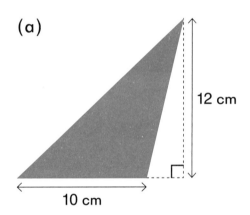

(b)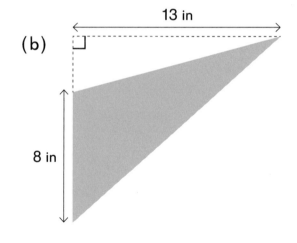

(c)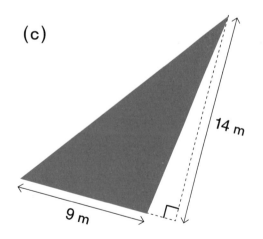

(d)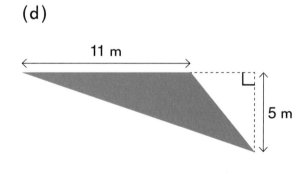

3) Find the area of each triangle.

(a)

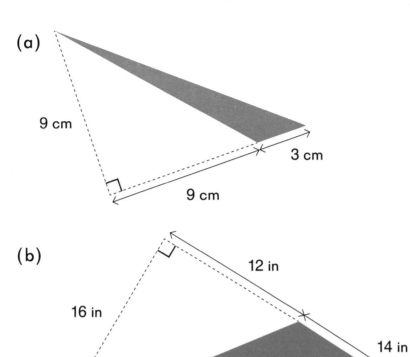

(b)

4) Triangles ABC, ABD, and ABE share the same base AB. Each triangle has a vertex on the same line parallel to the base. Find the area of each triangle. What do you notice about the area of each triangle?

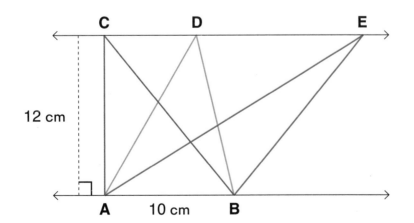

The triangles all have the same height.

5 Triangles EFG and EHG share the same base, EG. The height of Triangle EHG is twice as long as the height of Triangle EFG. Find the area of each triangle in square units and compare the areas.

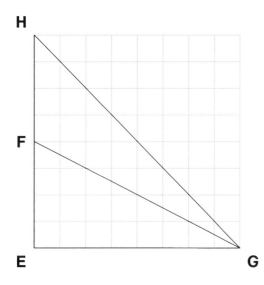

6 Triangles JKL and JKM share the same height, JK. The base of Triangle JKM is three times as long as the base of Triangle JKL. Find the area of each triangle and compare the areas.

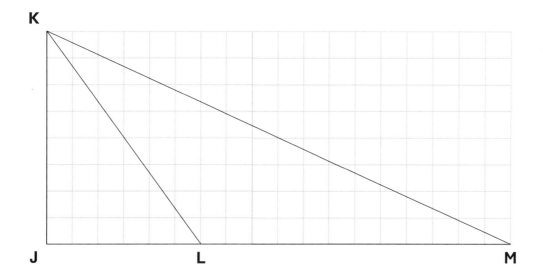

Exercise 5 • page 171

Lesson 6
Area of Complex Figures

Think

Find the area of Figure ABCD.

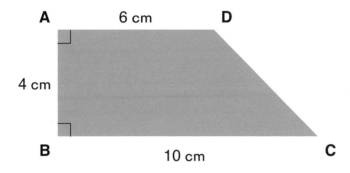

Learn

Method 1

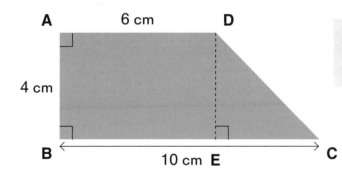

Area of Rectangle ABED = 6 × 4 = 24 cm²

Area of Triangle CDE = $\frac{1}{2}$ × (10 − 6) × 4

$= \frac{1}{2} × 4 × 4$

$= 8$ cm²

Area of Figure ABCD = 24 + 8 = 32 cm²

I cut the figure into a rectangle and a triangle and added the areas.

Method 2

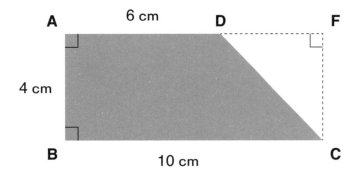

I made a larger rectangle and subtracted the area of triangle CDF.

Area of Rectangle ABCF = 10 × 4 = 40 cm²

Area of Triangle CDF = $\frac{1}{2}$ × (10 − 6) × 4 = 8 cm²

Area of Figure ABCD = 40 − 8 = 32 cm²

The area of Figure ABCD is _____ cm².

Do

1 Find the area of the shaded part of each rectangle.

(a)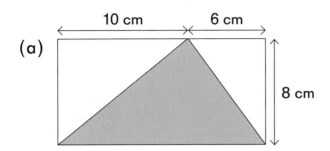

(b)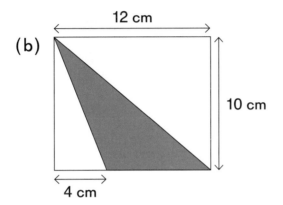

2 Find the area of the shaded part of each rectangle.

(a)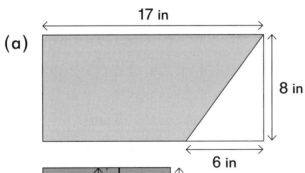

(b)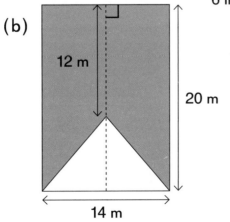

3 Find the area of figure EFGHI.

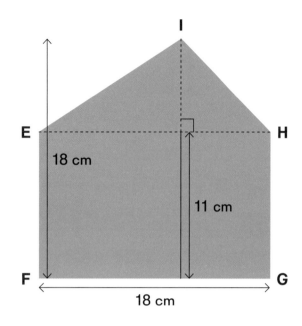

What is the height of Triangle EHI?

Area of Rectangle EFGH = ☐ cm²

Area of Triangle EHI = ☐ cm²

Area of Figure EFGHI = ☐ + ☐ = ☐ cm²

4 Find the area of each figure.

(a)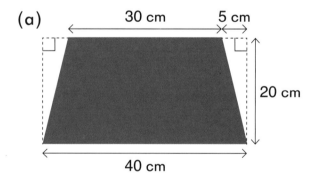

(b)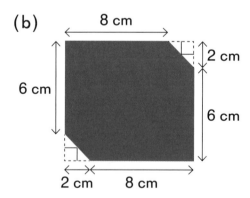

Exercise 6 • page 175

Lesson 7
Practice B

1 Find the area of each triangle.

(a)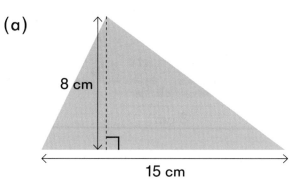

(b)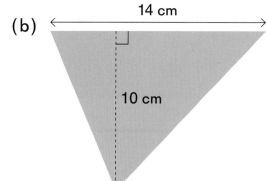

(c)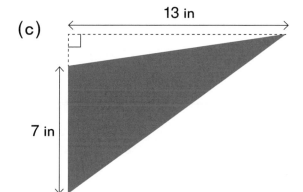

(d)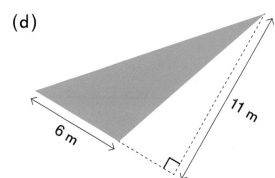

2 The perimeter of the triangle below is 50 cm. Find the area of the triangle.

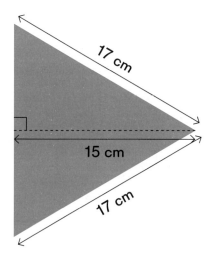

③ Find the area of each figure.

(a)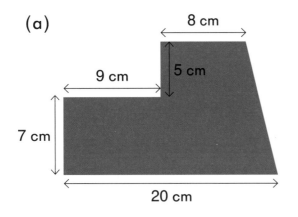

(b)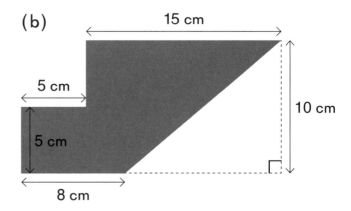

④ Find the shaded area.

(a)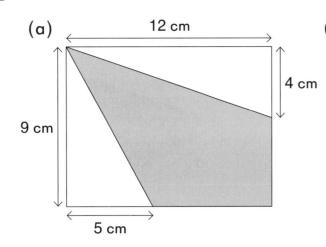

(b)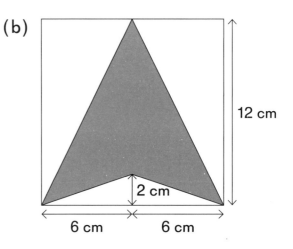

⑤ Find the area of each figure.

(a)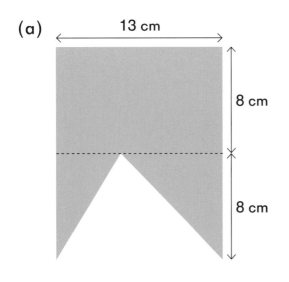

(b)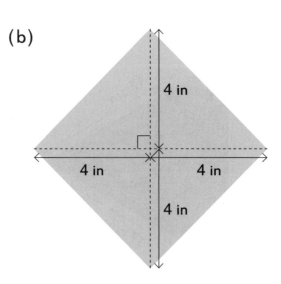

Exercise 7 • page 180

Chapter 8

Volume of Solid Figures

Figures A, B, and C are **solid figures** made with unit cubes.

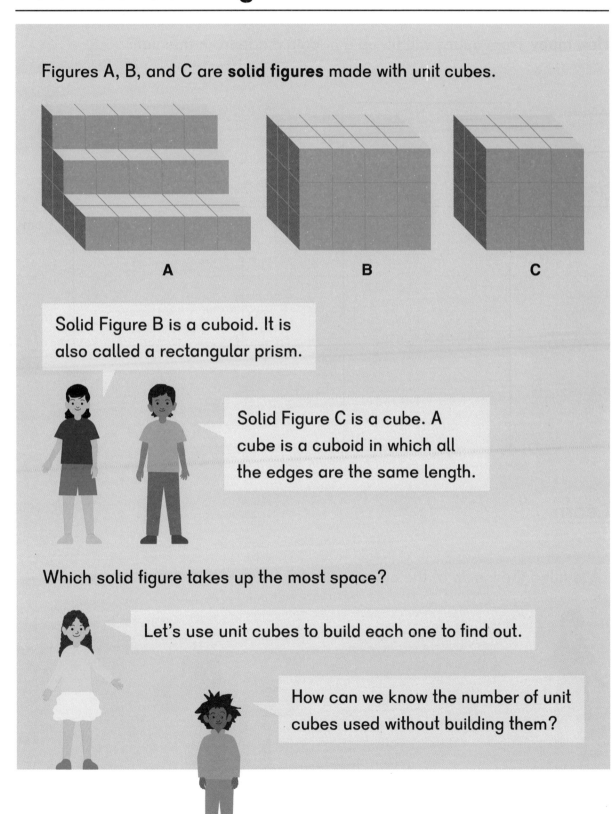

Solid Figure B is a cuboid. It is also called a rectangular prism.

Solid Figure C is a cube. A cube is a cuboid in which all the edges are the same length.

Which solid figure takes up the most space?

Let's use unit cubes to build each one to find out.

How can we know the number of unit cubes used without building them?

Think

How many 1-cm cubes will fill up a cuboid made from this net?

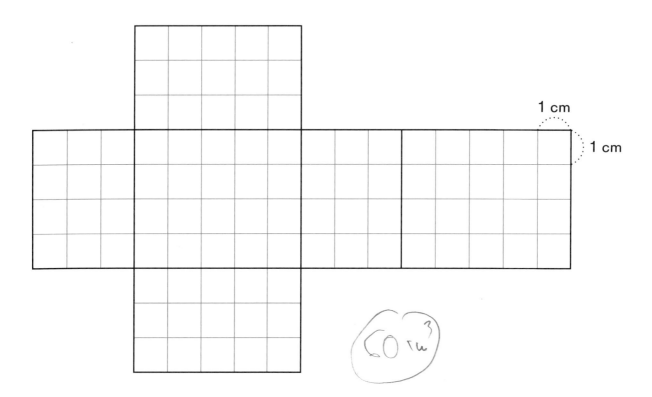

60 cu³

Learn

We filled the inside of the cuboid with 60 1-cm cubes.

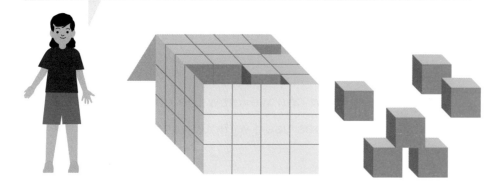

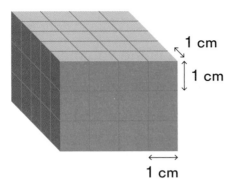

The amount of space a solid figure occupies is called the **volume** of the solid figure. We measure volume in **cubic units**. The volume of a unit cube is 1 cubic unit.

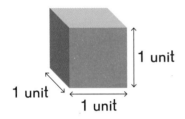

We are measuring the amount of three-dimensional space a solid figure takes up. We need units that take up three dimensions.

Some units of volume are **cubic centimeters** (cm^3), **cubic meters** (m^3), **cubic inches** (in^3), and **cubic feet** (ft^3).

We need sixty 1-cm cubes to fill the cuboid. The volume of the cuboid is 60 cm^3.

... solid figures below with 1-cm cubes. What is the volume of ... solid figure?

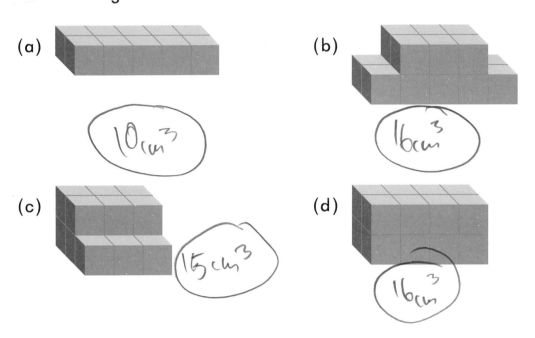

(a) 10 cm³

(b) 16 cm³

(c) 15 cm³

(d) 16 cm³

2 The solid figures below are made with unit cubes. What is the volume of each solid figure in cubic units?

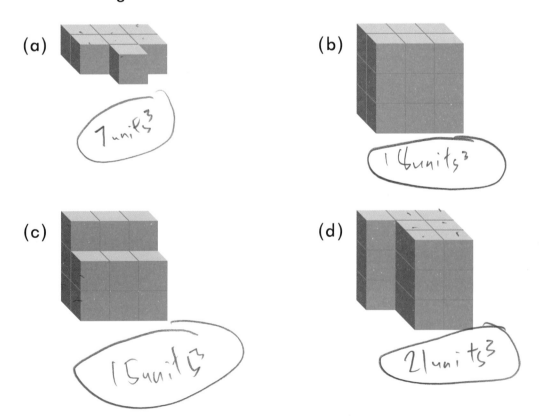

(a) 7 units³

(b) 16 units³

(c) 15 units³

(d) 21 units³

198   8-1 Cubic Units

3) The solid figures below were made with 1-cm cubes. Find the volume of each solid figure.

(a) Volume = 60 cm³

(b) Volume = 75 cm³

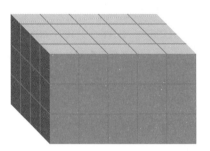

(c) Volume = 40 cm³

(d) Volume = 45 cm³

4) Make your own solid figure with unit cubes and ask a friend to find the volume.

Exercise 1 • page 185

Lesson 2
Volume of Cuboids

Think

How can we find the volume of the cuboid without building the whole shape?

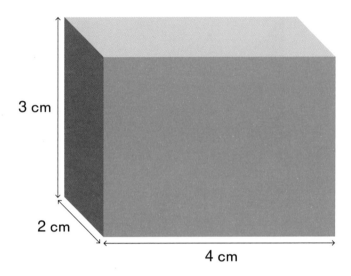

How can we figure out how many 1-cm cubes are needed to fill the cuboid?

How many 1-cm cubes do we need to build the bottom layer?

Learn

The number of 1-cm cubes on the bottom layer is 4 × 2 = 8.

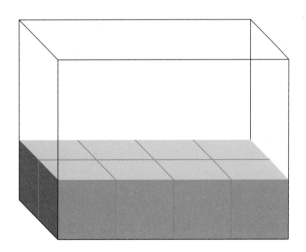

We will need three layers of cubes. Each layer has a volume of 8 cm³.

Volume of Cuboid = 4 × 2 × 3 = 8 × 3 = 24 cm³

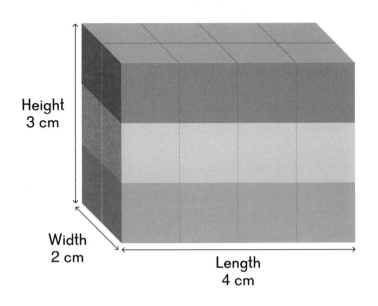

The base has an area that is equal to the number of cubes in the bottom layer. The height is the same value as the number of layers.

Volume of a Cuboid = Area of Base × Height
= Length × Width × Height

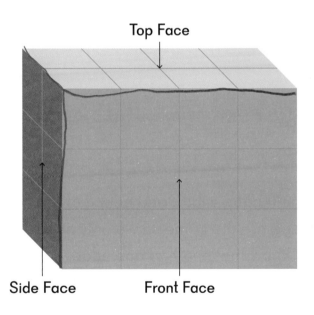

We can find the volume by multiplying the area of any face by the length of the edge perpendicular to the face.

Area of Top Face × Height = (4 × 2) × 3 = 24 cm³

Area of Side Face × Length = (2 × 3) × 4 = 24 cm³

Area of Front Face × Width = (4 × 3) × 2 = 24 cm³

Do

1 The cuboids below are made of 1-cm cubes. Find the volume of each cuboid.

(a)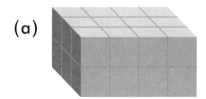

Length × Width × Height = 4 × 3 × 2 = 24 cm³

Area of Base × Height = (4 × 3) × 2 = 24 cm³

(b)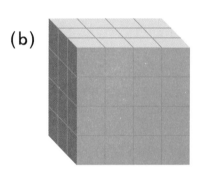

Volume = 3 × 4 × 4 = 46 cm³

2 Find the volume of the cube.

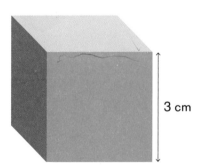

3 cm

The length, width, and height of the cube are the same so we only need to know the length of one edge.

Volume of Cube = 3 × 3 × 3 = 27 cm³

3 A box is partially filled with 1-cm cubes. What is the volume o

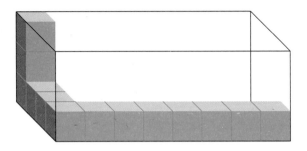

$3 \times 4 \times 8 = 96 \text{ cm}^3$

4 Find the volume of each cuboid.

(a)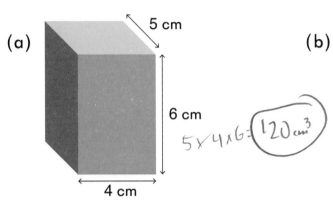

$5 \times 4 \times 6 = 120 \text{ cm}^3$

(b)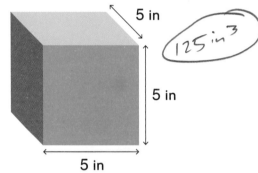

125 in^3

(c)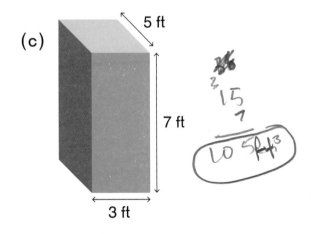

15
7
105 ft^3

(d)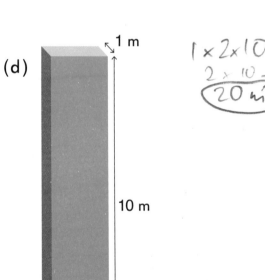

$1 \times 2 \times 10$
2×10
20 m^3

Make sure to express each answer with the appropriate cubic unit.

8-2 Volume of Cuboids

5 Find the volume of each cuboid.

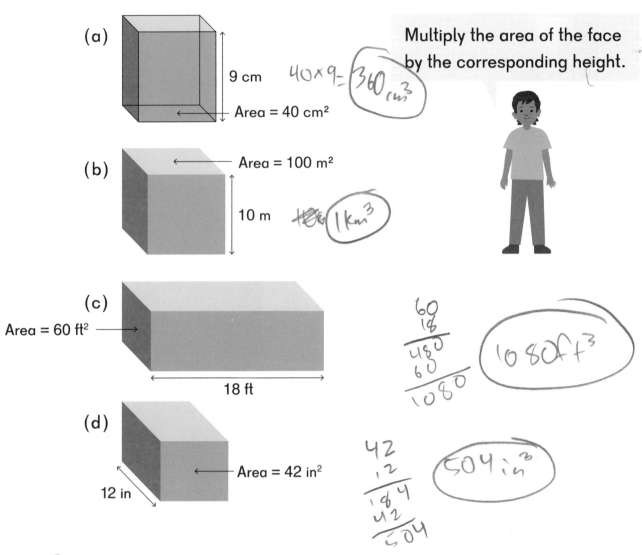

(a) 9 cm, Area = 40 cm² — 40×9 = 360 cm³

Multiply the area of the face by the corresponding height.

(b) Area = 100 m², 10 m — 1 km³

(c) Area = 60 ft², 18 ft — 1080 ft³

(d) 12 in, Area = 42 in² — 504 in³

6 Mei wants to choose the box with the greatest volume to store her stuffed animals. Which box should she choose?

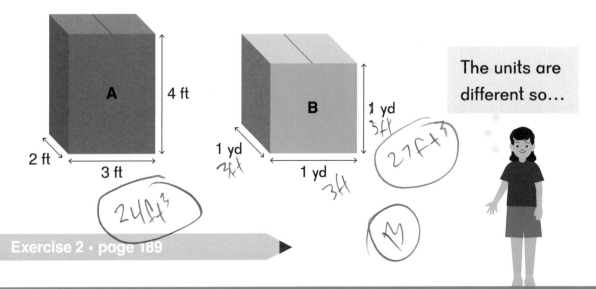

A: 2 ft × 3 ft × 4 ft — 24 ft³

B: 1 yd × 1 yd × 1 yd (3 ft) — 27 ft³

The units are different so...

Exercise 2 • page 189

Lesson 3
Finding the Length of an Edge

Think

The volume of a cuboid is 24 cm³. The length of the cuboid is 4 cm and the width is 2 cm. What is the height of the cuboid?

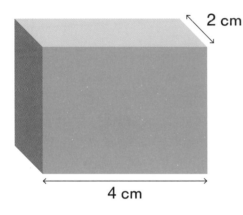

Learn

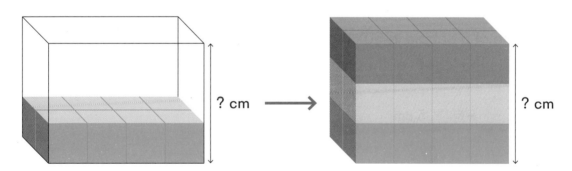

4 × 2 × ? = 24

8 × ? = 24

? = $\frac{24}{8}$ = 3

Height = 3 cm

The cuboid has a height of ___3___ cm.

To find the length of an edge of a cuboid, divide the volume by the product of the lengths of the other two edges.

1. The volume of a cuboid is 36 cm³. It has a length of 4 cm and a height of 3 cm. What is the width of the cuboid?

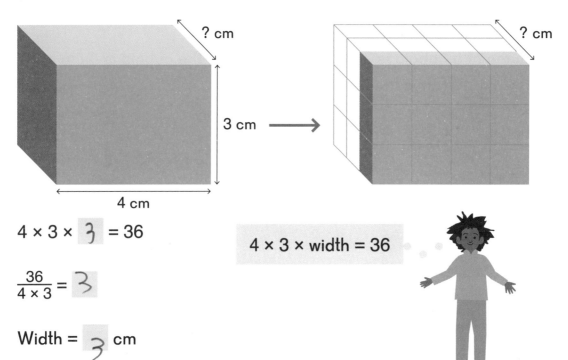

4 × 3 × 3 = 36

$\frac{36}{4 \times 3}$ = 3

Width = 3 cm

4 × 3 × width = 36

2. The volume of a cuboid is 60 cm³. It has a width of 5 cm and a height of 4 cm. What is the length of the cuboid?

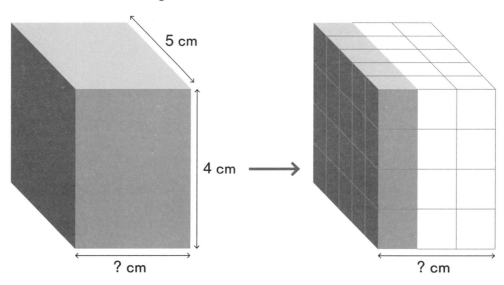

Length = 3 cm

5 × 4 × length = 60, so length = $\frac{60}{5 \times 4}$.

8-3 Finding the Length of an Edge

3 The volume of this cuboid is 40 cm³. The area of the base is What is the height?

10 × 4 = 40

$\frac{40}{10}$ = 4

Height = 4 cm

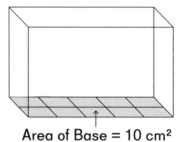
Area of Base = 10 cm²

4 Find the length of the unknown edge of each cuboid.

(a) Volume = 168 in³

AB = 14 in

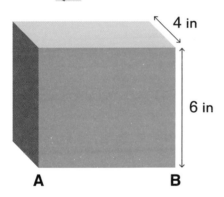

(b) Volume = 343 m³

CD = 7 m

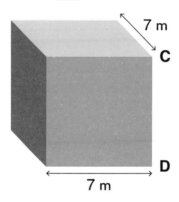

(c) Volume = 360 ft³

EF = 8 ft

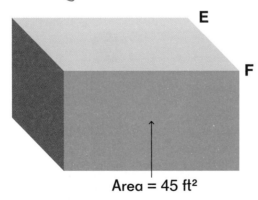

(d) Volume = 140 yd³

GH = 5 yd

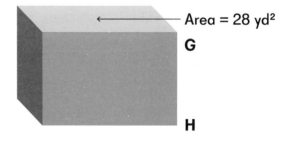
Area = 28 yd²

5. The volume of a cube is 125 cm³. What is the length of one edge?

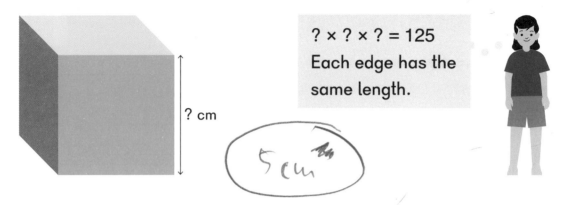

? × ? × ? = 125
Each edge has the same length.

Handwritten answer: 5 cm

6. The volume of the cuboid below is 24 cm³. What is the area of the base?

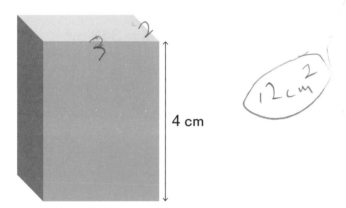

Handwritten answer: 12 cm²

7. The cube-shaped box has a volume of 1 m³. How many 1-cm cubes can fit in the box?

1 m = 100 cm

Handwritten work: 1,000,000; 1 million cubes

Exercise 3 • page 192

8-3 Finding the Length of an Edge

Lesson 4
Practice A

P 4

1 The solid figures below are built with 1-cm cubes. Find the volume of each solid figure.

(a)

(b)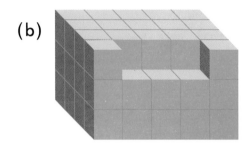

(c)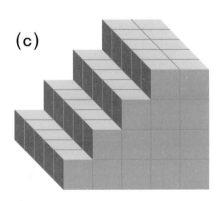

(d)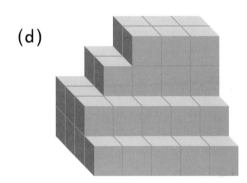

2 Find the volume of each cuboid.

(a)

(b)

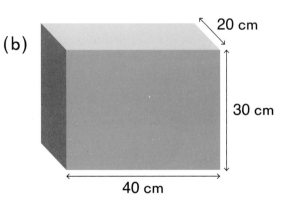

3 Find the volume of each cube.

(a)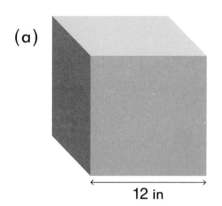
12 in

(b)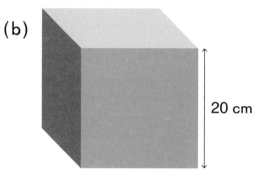
20 cm

(c)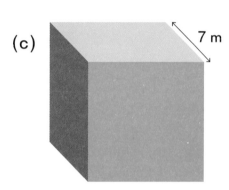
7 m

(d)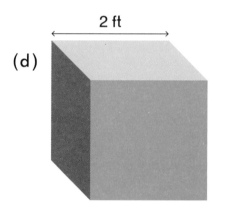
2 ft

4 Find the volume of each cuboid.

(a)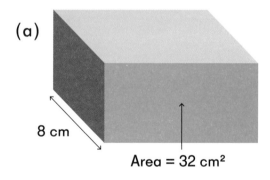
8 cm
Area = 32 cm²

(b)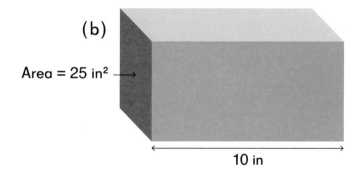
Area = 25 in²
10 in

5. Find the length of the missing edge.

(a) Volume = 150 cm³

Height = ☐ cm

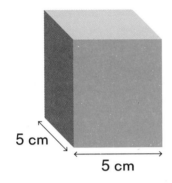

(b) Volume = 84 m³

Length = ☐ m

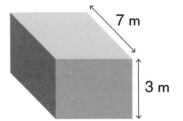

6. Find the length of an edge of each cube.

(a) Volume = 27 in³

Edge = ☐ in

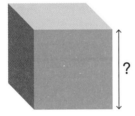

(b) Volume = 64 cm³

Edge = ☐ cm

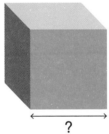

7. A box has a length of 18 cm, a width of 8 cm, and a height of 15 cm.

(a) What is the volume of the box?

(b) How many cubes with an edge of 2 cm will fit in the box?

Exercise 4 • page 195

5
Volume of Complex Shapes

Find the volume of the solid figure.

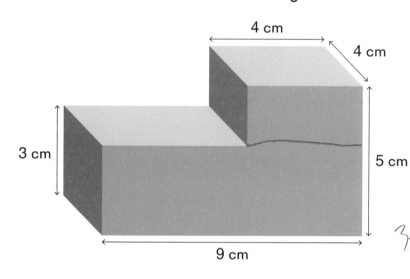

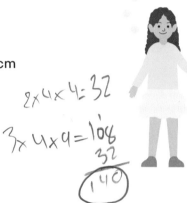

How can we use cuboids to find the volume?

Learn

Method 1

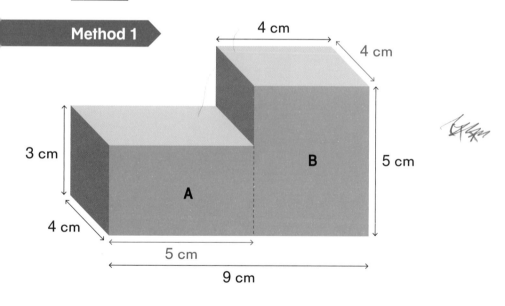

Volume of Cuboid A = 5 × 4 × 3 = 60 cm³

Volume of Cuboid B = 4 × 4 × 5 = 80 cm³

Volume of the solid figure = 60 + 80 = 140 cm³

Method 2

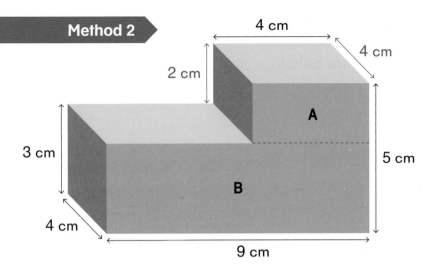

Volume of Cuboid A = 4 × 4 × 2 = 32 cm³

Volume of Cuboid B = 9 × 4 × 3 = 108 cm³

Volume of the solid figure = 32 + 108 = 140 cm³

Method 3

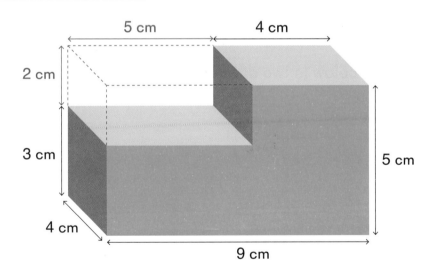

Volume of the large cuboid = 9 × 4 × 5 = 180 cm³

Volume of cut-away cuboid = 5 × 4 × 2 = 40 cm³

Volume of the solid figure = 180 − 40 = 140 cm³

The solid has a volume of __140__ cm³.

Do

① Use unit cubes to build the solid figure below.

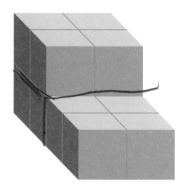

(a) In what different ways can you divide the solid figure into 2 cuboids?

hor ⟷ vert cut away

(b) What is the fewest number of cubes that need to be added to change the figure into a cuboid?

4

(c) What is the volume of the figure?

12

② Find the volume of the solid figure below in two different ways.

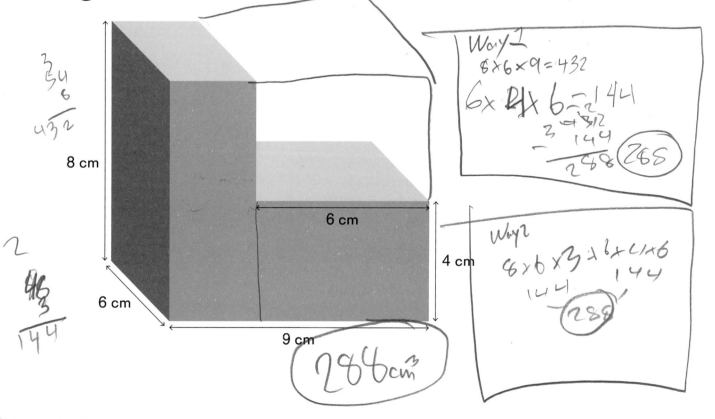

288 cm³

Way 1
8×6×9 = 432
6×4×6 = 144
288

Way 2
8×6×3 + 6×4×6
144   144
288

8-5 Volume of Complex Shapes

3 Find the volume of each solid figure.

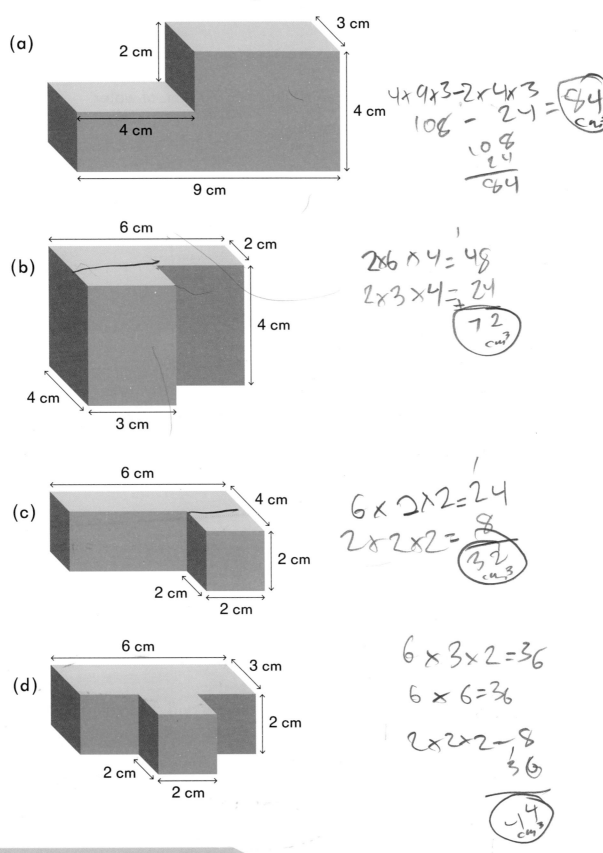

(a) 4 × 9 × 3 − 2 × 4 × 3 = 84 cm³
108 − 24 = 84
108
 24

  84

(b) 2 × 6 × 4 = 48
2 × 3 × 4 = 24
72 cm³

(c) 6 × 2 × 2 = 24
2 × 2 × 2 = 8
32 cm³

(d) 6 × 3 × 2 = 36
6 × 6 = 36
2 × 2 × 2 = 8
36
 8
---
44 cm³

Exercise 5 • page 199

8-5 Volume of Complex Shapes

Lesson 6
Volume and Capacity — Part 1

Think

A 1-liter cube-shaped container is filled with 700 mL of water.

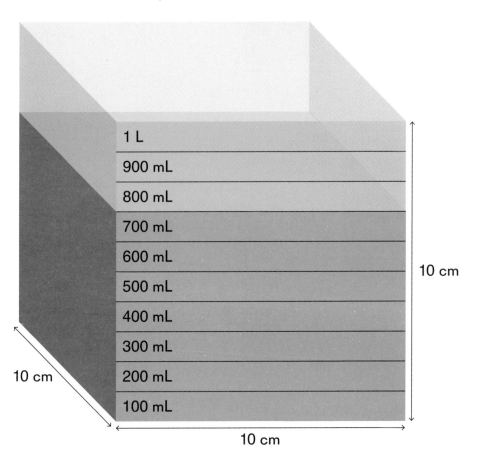

(a) What is the capacity of the container in milliliters?

(b) What is the volume of the container in cubic centimeters?

(c) What is the relationship between milliliters and cubic centimeters?

(d) What is the volume of the empty space in the container?

Learn

(a) Capacity of Container = 1 L = 1,000 mL

(b) Volume of Container = 10 × 10 × 10 = 1,000 cm³

(c) 1 mL = 1 cm³

1,000 mL = 1,000 cm³

(d)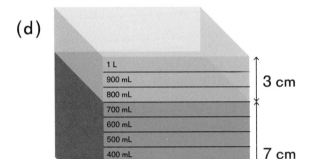

Method 1

Volume of container = 1,000 cm³

Volume of water = 700 mL = 700 cm³

Volume of empty space = 1,000 − 700 = 300 cm³

We would need 300 mL more water to fill the container to capacity.

Method 2

Height of empty space = 10 − 7 = 3 cm

Volume of empty space = 3 × 10 × 10 = 300 cm³

The volume of the empty space is __300__ cm³.

Do

1 Write each amount in cubic centimeters.

(a) 750 mL = 750 cm³

(b) 2 L = 2000 cm³

(c) 1 L 300 mL = 1300 cm³

(d) 3 L 25 mL = 3025 cm³

2 Write each amount in liters and milliliters.

(a) 650 cm³ = 650 mL

(b) 4,000 cm³ = 4 L

(c) 1,850 cm³ = 1 L 850 mL

(d) 2,005 cm³ = 2 L 5 mL

3 A cube-shaped container measures 15 cm by 15 cm by 15 cm. It is filled with 3 L of water.

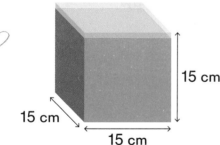

3 L = ? cm³

(a) What is the capacity of the container in milliliters?

(b) How many more milliliters of water is needed for the container to be completely full?

4. A cuboid-shaped fish tank measures 30 cm by 15 cm by 20 cm. It is filled with water to a height of 10 cm.

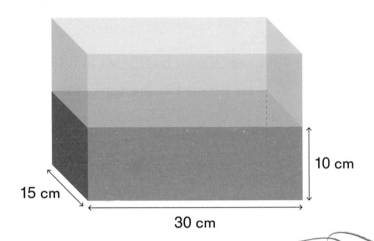

(a) What is the volume of water in the fish tank in cubic centimeters?

 4500 cm³

(b) What is the volume of water in liters and milliliters?

 4 L 500 mL

(c) What is the capacity of the fish tank in liters and milliliters?

 9 L

(d) What is the volume of the empty space in the tank in cubic centimeters?

 4500 cm³

5. The area of the bottom side of a cuboid-shaped fish tank is 100 cm². There is $1\frac{1}{2}$ L of water in the tank. What is the height of the water?

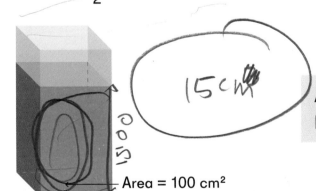

15 cm

Area of Base × Height = Volume
Height = Volume ÷ Area of Base

Exercise 6 • page 202

Lesson 7
Volume and Capacity — Part 2

Think

A cuboid-shaped tank with a base that measures 30 cm by 20 cm is filled with water to a height of 10 cm. After a rock was placed in the tank, the height of the water rose to 12 cm. What is the volume of the rock?

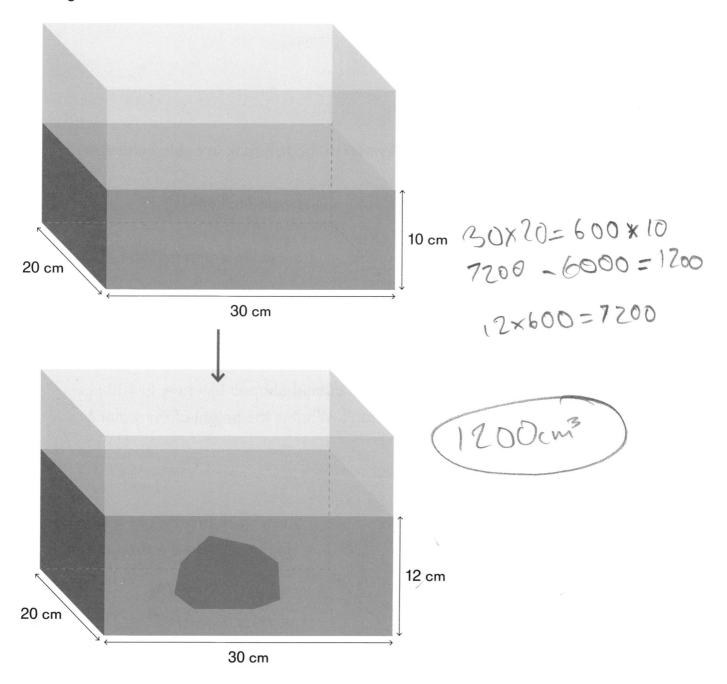

30 × 20 = 600 × 10
7200 − 6000 = 1200
12 × 600 = 7200

1200 cm³

Learn

Method 1

Volume of the water in the tank before the rock was put in
= 20 × 30 × 10 = 6,000 cm³

Volume of the water and the rock together = 30 × 20 × 12 = 7,200 cm³

Volume of the rock = 7,200 − 6,000 = 1,200 cm³

Method 2

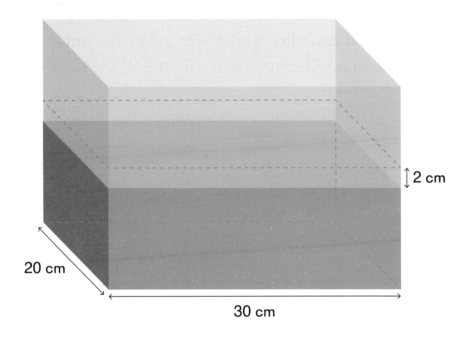

Difference between the height of water before and after the rock was put in
= 12 − 10 = 2 cm

Volume of rock = 30 × 20 × 2 = 1,200 cm³

The rock has a volume of __1200__ cm³.

Do

1 A measuring cylinder has 57 mL of water. After Alex dropped some pebbles in it, the water level rose to 82 mL. What is the volume of the pebbles in cubic centimeters?

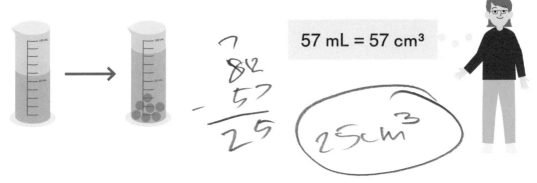

57 mL = 57 cm³

82 − 57 = 25

25 cm³

2 A rectangular tank that has a length of 40 cm and a width of 30 cm is filled with water to a height of 18 cm. After a brick was put in, the water rose to a height of 23 cm. What is the volume of the brick?

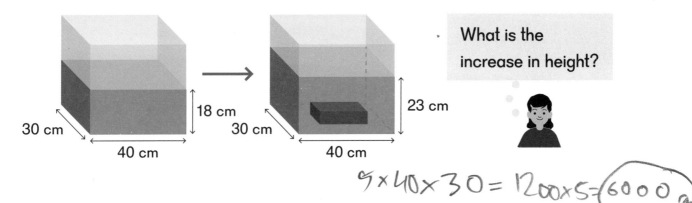

What is the increase in height?

5 × 40 × 30 = 1200 × 5 = 6000 cm³

3 A rectangular tank with a length of 50 in and a width of 30 in has some water and a brick in it. After the brick was taken out, the water level decreased by 5 in. What is the volume of the brick?

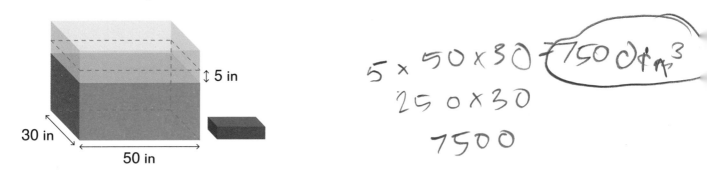

5 × 50 × 30 = 7500 in³
250 × 30
7500

Exercise 7 • page 205

Lesson 8
Practice B

1 Find the volume of each solid figure.

(a)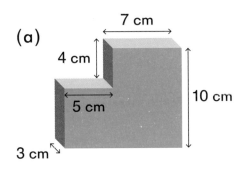

(b)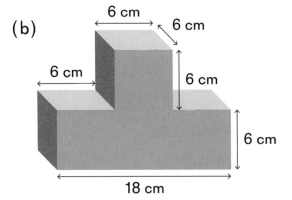

2 A rectangular tank with a length of 30 cm and a width of 20 cm contained water to a height of 12 cm. A metal can with a volume of 1,500 cm³ is placed in the tank. What is the new height of water in the tank in centimeters?

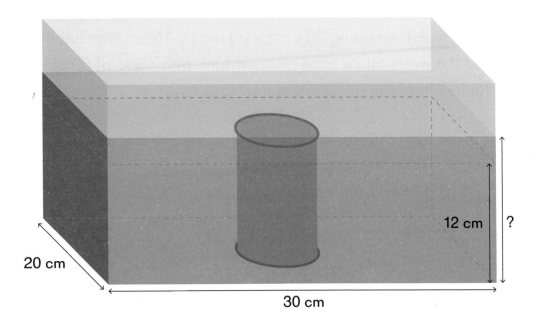

3 A rectangular container with a length of 10 cm and a width of 10 cm is filled with water to a height of 8 cm. After 8 identical marbles were added the height of the water rose to 12 cm. What is the volume of one marble?

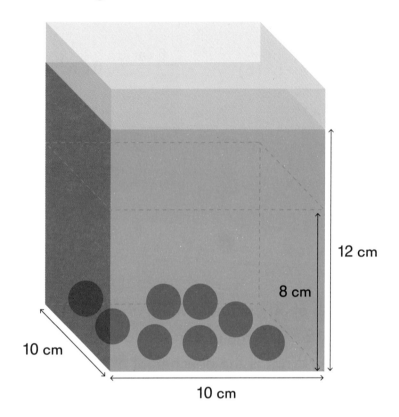

4 A rectangular tank measuring 30 cm by 20 cm by 15 cm was full of water. Dion poured the water into a cube-shaped tank with sides of 30 cm. What is the height of the water in the new tank?

5 A metal object with a volume of 2,500 cm³ was placed in a tank with a length of 20 cm. The water rose 2 cm. What is the width of the tank in centimeters?

6 A large water tank 2 m by 1 m by $1\frac{1}{2}$ m is filled with water to a height of $\frac{1}{2}$ m. How many liters of water is needed to fill the rest of the tank?

Exercise 8 • page 208

Review 2

1 Find the values. Express each answer in simplest form.

(a) $4\frac{5}{6} + 4\frac{3}{5}$

(b) $9\frac{7}{10} - 4\frac{5}{6}$

2 Estimate and then find the actual values, expressed in simplest form.

(a) 432×65

(b) $8{,}671 \times 24$

(c) $78 \times 7{,}342$

(d) $8{,}956 \div 27$

(e) $3{,}816 \div 42$

(f) $98{,}532 \div 24$

3 Find the values. Express each answer in simplest form.

(a) $12 \times \frac{3}{8}$

(b) $14 \times \frac{5}{7}$

(c) $\frac{8}{3} \times 24$

(d) $\frac{1}{9} \times \frac{1}{8}$

(e) $\frac{7}{10} \times \frac{5}{21}$

(f) $\frac{8}{5} \times \frac{10}{4}$

(g) $5\frac{5}{6} \times \frac{3}{5}$

(h) $3\frac{2}{5} \times 3\frac{1}{3}$

(i) $1\frac{4}{5} \times 2\frac{2}{9}$

(j) $\frac{3}{4} \times \frac{5}{7} \times \frac{4}{3} \times \frac{7}{5}$

(k) $\frac{2}{3} \times 9 + \frac{2}{3} \times 3$

(l) $15 - 8 \times \frac{1}{2} + 2\frac{4}{5}$

(m) $3\frac{6}{7} \times 8 - 1\frac{1}{4} \times 8$

(n) $\frac{2}{4} \times (3\frac{1}{2} - 2\frac{2}{3}) + 6\frac{1}{2}$

4 Find the missing numbers.

(a) $\frac{7}{8} \times 1\frac{1}{7} =$ ☐

(b) $\frac{4}{5} \times$ ☐ $= 1$

(c) ☐ $\times 7 = 1$

(d) $1 = 4\frac{2}{3} \times$ ☐

5 Find the missing numbers.

(a) $\frac{3}{5}$ h = ☐ min

(b) $\frac{7}{10}$ km = ☐ m

(c) $3\frac{1}{4}$ gal = ☐ qt

(d) $5\frac{3}{4}$ ft = ☐ in

(e) $3\frac{1}{4}$ L = ☐ mL

(f) $2\frac{3}{5}$ m = ☐ cm

6 Find the values. Express each answer in simplest form.

(a) $\frac{1}{8} \div 5$

(b) $\frac{1}{3} \div 12$

(c) $\frac{9}{10} \div 3$

(d) $\frac{3}{5} \div 6$

(e) $\frac{15}{9} \div 5$

(f) $\frac{1}{9} \div 4$

(g) $6 \div \frac{1}{10}$

(h) $12 \div \frac{1}{4}$

(i) $6 \div \frac{3}{8}$

(j) $9 \div \frac{3}{7}$

(k) $\frac{5}{7} \div 10 \times \frac{20}{3}$

(l) $\frac{3}{5} \times 10 \div \frac{1}{3}$

(m) $4\frac{3}{4} - 2 \div \frac{1}{2}$

(n) $7 \div (\frac{5}{6} - \frac{2}{3}) \times 75$

7 Gina spent $\frac{2}{3}$ of her savings on a car. The car cost $2,916. How much were her savings to begin with?

8 Barry had a rope that was 36 ft long. He used $\frac{3}{4}$ of it to make a swing and $\frac{1}{3}$ of the remainder to repair a fence. How much rope does he have left?

9 A carpenter cut a 21-ft long board into pieces that are each $\frac{3}{4}$ ft long. How many $\frac{3}{4}$-ft long pieces are there?

10 $\frac{4}{5}$ kg of pumpkin seeds are placed equally into 8 bags. How many kilograms of pumpkin seeds are in each bag?

11 It takes $\frac{2}{3}$ cups of flour to make 4 dinner rolls. How much flour does it take to make 1 dinner roll?

12 Sasha had $400. She spent $\frac{3}{8}$ of it on a dress and $\frac{4}{5}$ of the remainder on shoes. She spent what was left on a coat. How much did the coat cost?

13 $\frac{2}{3}$ of the coins in Caleb's coin collection are gold coins, $\frac{1}{3}$ of the remainder are silver coins, and the rest are copper coins. He has 100 copper coins. How many coins does he have in his collection?

14 $\frac{3}{7}$ of Eli's stamps were French stamps and the rest were Spanish stamps. He gave $\frac{1}{3}$ of his Spanish stamps to Mia. What fraction of the stamps that he started with does he have left?

15 An employee at a market has 27 lb of grapes. She puts $\frac{2}{3}$ of the grapes in a bin and packs the remainder of the grapes into bags. Each bag has $\frac{3}{4}$ lb of grapes. How many bags of grapes are there?

16 There were 8,896 books in two libraries. After $\frac{1}{4}$ of the books were transferred from Library A to Library B, there were $\frac{3}{5}$ as many books in Library A as in Library B. How many books are now in Library B?

17 Find the area of each figure.

(a)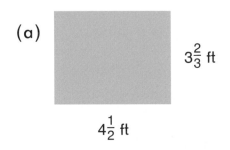

(b)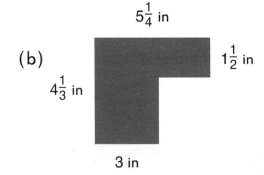

(c)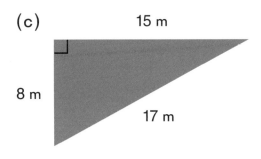

(d)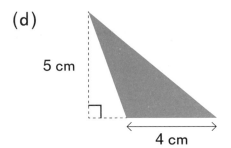

(e)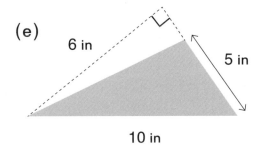

(f)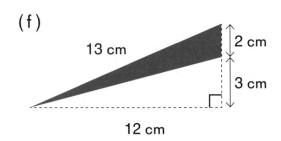

Review 2

18 Find the shaded area of each figure.

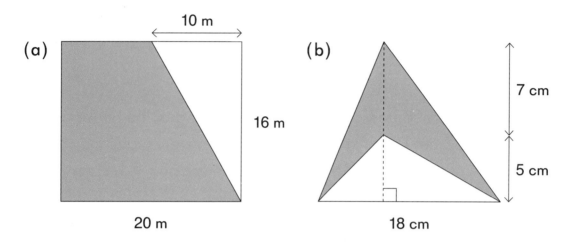

19 Find the volume of each solid figure. Figure (a) is a cube.

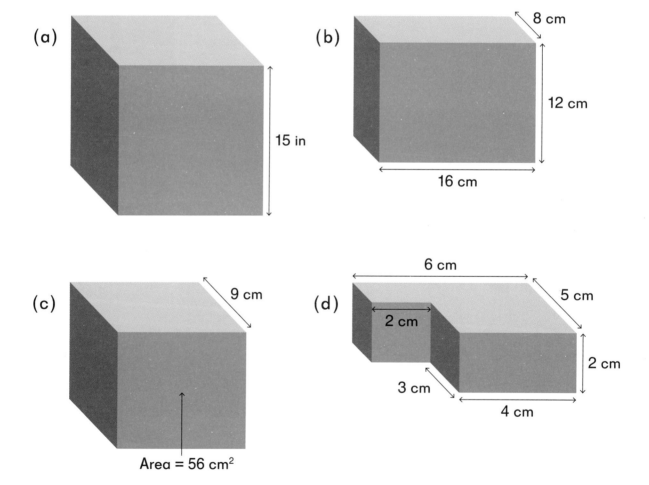

20 The cuboid has a volume of 960 cm³. Find the length of the Edge AB.

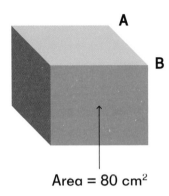

Area = 80 cm²

21 A rectangular container that measures 24 cm by 12 cm by 20 cm is filled to the top with water. 2 L of water are emptied from the container. How much water is left in the container?

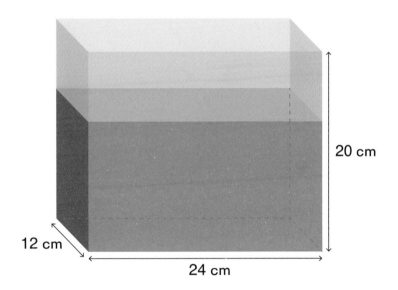

22 A rectangular fish tank has a length of 48 in, width of 12 in, and height of 16 in. It is filled with water to a height of 10 in. After some rocks are placed in the tank the water level rose to a height of 14 in. What is the volume of the rocks?

Exercise 9 • page 212

Blank